FOR FUN AND PROFIT

# HISTORY OF COMPUTING

William Aspray and Thomas J. Misa, editors

# FOR FUN AND PROFIT

**A History of the
Free and Open Source Software
Revolution**

CHRISTOPHER TOZZI
Foreword by Jonathan Zittrain

The MIT Press
Cambridge, Massachusetts
London, England

This book was set in Adobe Garamond and Berthold Akzidenz Grotesk by the MIT Press. Printed and bound in the United States of America.

Library of Congress Cataloging-in-Publication Data is available.

ISBN: 978-0-262-03647-4

10   9   8   7   6   5   4   3   2   1

# Contents

# Foreword

Jonathan Zittrain

THE IDEALS of the free software movement are fundamental to our relationship to technology. Their goals are for us to be able to learn from and alter the increasingly baroque and pervasive code that shapes our lives. By learning about and refining that code, we can better understand and affect the world around us, inspiring the kinds of self-reliance and self-realization that are some of the most important aspects[1] of human flourishing.

Defending these ideals also could help prevent us from being unduly corralled and cabined. Arthur C. Clarke observed that any sufficiently advanced technology is indistinguishable from magic,[2] and when something is magic, it concentrates power in the hands of magicians and those who employ or regulate them.

Although the ideals of the free software movement are timeless, realizing them has become increasingly difficult to achieve. The environment for free software in the 1980s and 1990s benefited from the work that had been done by academia—and according to its values—in the computing space. As you will learn in this book, the pioneering (and ultimately free) Unix software operating system arose from the singular situation of Bell Laboratories, a corporate source that, due to a consent decree involving the Bell system, was at first constrained in pro-prietizing it. Academics were ready to extend and later rebuild the entire system to keep it open.

Today, most corporate sources of operating platforms, whether for the gadgets we buy or the cloud they connect with, are under no such constraint. Some players possess a war chest and market dominance sufficient to introduce new code without relying much on the old—and its attendant free licensing arrangements.

Ironically, the openness that made the Internet possible and all-embracing has meant that code can be run at a distance over the network as readily as it can on the device itself. The rise of cloud computing should not affect functionality, but the predicate act for reinforcing the original core free software licenses—if you copy the code to give it to someone else, including a customer, you must pass along the tools to read it and change it—is no longer part of using and profiting from the code. Running that software now does not need that transfer, which means it will not trigger the freedom-enhancing requirements of most free software licenses.

Today we confront a world in which the software products that assist us, make decisions about us, and render our books and other content are both more powerful and more obscure than ever. The many-faceted story you are about to read is not only a rollicking tale of fierce personalities and human conflicts around passionately held views but also a source of inspiration for how to deal with our increasingly algorithmicized world.

# Acknowledgments

THIS BOOK EXISTS because many generous people took chances on me by offering gifts of time, patience, and material goods.

My grandparents, Robert and Dorothy Schatzle, spent a fortune to buy my first computer, an IBM PS/2, when I was in second grade, hoping to stoke a new creative outlet. Andy Tozzi, my uncle, volunteered his Saturday mornings to take me to computer fairs and teach me about Windows 3.1. Many years later, he burned my first GNU/Linux installation CD. (It was SUSE.) Another grandmother, Mary Ann Tozzi, taught me how to write emails when she had much more important things to do with her time.

In college and graduate school, Mark Sanford and Joe Stone gave me student jobs administering GNU/Linux servers, even though nothing on my transcript at the time suggested that I deserved them. Joe Panettieri offered me my first paid writing gig, as a contributor to a now-defunct blog about Ubuntu, when I had zero publications to my name. Charlene O'Hanlon, who edited my work, patiently helped to improve my writing—and to clean up messes when I accidentally broke news embargoes.

David A. Bell, who supervised my PhD dissertation in French history, encouraged me to work on this project, even though I worried that it might look strange on my curriculum vitae. He reminded me that academia would be a boring place indeed if professors were afraid to study whatever happens to interest them.

The anonymous readers of my manuscript for the MIT Press offered excellent and insightful feedback, which helped considerably to sharpen the content of this book. I am grateful, too, to the staff of the MIT Press, especially Katie Helke, for making the publication of this work a smooth process.

Many readers of my articles about software history at *The VAR Guy* offered comments that helped to hone the ideas presented in this book. One reader, Art Protin, who was a stranger to me, took the time to read and comment on a complete draft of the book. For that, I am grateful indeed. Attendees of the Columbia Area Linux Users' Group, who invited me to speak about this project, also provided much valuable feedback on this work.

Dave Schneider, who offered an opportunity to publish some material related to this book in *IEEE Spectrum* magazine, delivered brilliant editing and guidance, which benefited the project as a whole.

Kate Sohasky took the greatest chance and gave the greatest gift of all. She married me while I was in the middle of writing this book. (She was a Windows 7 user at the time but has since converted to Ubuntu.) The only thing that makes me happier than having married her is looking forward to the gifts we will share in the future.

# INTRODUCTION

"MAN IS BORN FREE, and everywhere he is in chains."[1] So lamented Jean-Jacques Rousseau in the opening lines of his 1762 treatise, *The Social Contract*. The phrase proved to be one of the most enduring criticisms of political and social inequality in prerevolutionary Europe. It helped to inspire the French revolutionary movement that, starting a decade after Rousseau's death in 1778, undid those chains and—in the minds of the revolutionaries, at least—tried to restore the equality between men that they imagined to have reigned supreme when nature, rather than human contrivance, governed society.

Had Rousseau lived two centuries later, he might have made a similar observation about software: "Code was born free, and everywhere it is in chains." The early 1980s marked the start of an era in which most programmers ceased to share code freely—despite having done so during the first decades of software development, before commercial interests transformed the early world of computing.

A computer-age Rousseau might have proved to be no less of a prescient philosopher than the real-life Enlightenment version. Just as the 1780s witnessed the start of a revolution against inequality in social and political affairs, so did the 1980s inaugurate a similar revolt that aimed to restore software freedom. By the end of the 1990s, the revolution had succeeded. In many places, code was free again.

*   *   *

To discuss eighteenth-century philosophers in tandem with twentieth-century technology may seem like nothing more than the basis for a bad science fiction novel. Yet the connection here is neither hackneyed nor contrived. Rousseau and his rabble-rousing contemporaries had much in common with a group of programmers who came of age near the end of the Cold War. Each of these factions helped to launch revolutions that changed the world.

They were very different sorts of revolutions. The French Revolution of 1789 sparked brutal internecine violence and decades of warfare. The free and open source software revolution that began in the 1980s never killed anyone. But both revolutions were radical and revisionist, and the effects of the latter revolution could be said to be as profound as those of the political upheavals of the revolutionary age. Just as politics and modes of social organization would not look like they do today if the eighteenth-century revolutions had not happened, so too would the way millions of people now use computers be wildly different if there had never been a revolution surrounding computer code.

That commonality is not a coincidence. The free and open source software (FOSS) revolution drew on the same ideological traditions as more violent revolutions that preceded it did. At times, the software revolutionaries even looked to the "revolutionary script" inaugurated by political revolutionaries centuries earlier for inspiration.[2]

This book explores the revolution that birthed free and open source software. It traces the revolution's origins. It narrates its different stages. It explains its unpredictable arc and unexpected outcomes. And it looks into the future to predict how FOSS will continue to evolve.

## WHAT IS FOSS?

For several reasons, it is not easy to define what the term *free and open source software* (FOSS) means for the purposes of this book. First of all, the term *free software* is ambiguous. Sometimes it refers to software that may be legally copied without cost. It also can describe software whose source code is governed by particular licenses designed to ensure that anyone can view it. Some programs described as free software have both of these qualities.

Meanwhile, there is no single definition for the term *open source*. The FOSS community generally uses this phrase to describe programs whose source code is publicly available under certain licenses, as opposed to other programs that it associates with free software. In general, the difference revolves around the extent to which the licenses ensure that derivative works retain the same copying permissions as those granted by the original license. Yet there is no universal consensus regarding which licenses make software open source and which licenses apply to free software.

Making matters more confusing is the fact that although there is a certain degree of interchangeability between the phrases *free software* and *open source software*, failing to distinguish between them to the liking of different groups of programmers can touch major nerves. Some developers are loathe to see their work labeled open source software when they consider it instead to be free software and vice versa. Journalists and marketers rarely distinguish the two terms effectively and sometimes fail to appreciate the differences between FOSS and software that merely costs no money, which does not help. Moreover, the phrase *open source software* did not exist before 1998, which makes it anachronistic to use the term to describe code that was developed before that date, even though many people frequently do when discussing historical events.

In addition, the meaning of the word *open* itself in reference to software is rife with ambiguity. There is usually little disagreement about what constitutes closed source software. The term *closed source software* refers to programs whose source code is not publicly shared. In contrast, "open" code takes many forms and means different things to different people. Some contend that sharing source code when requested makes it open. Others argue that allowing third parties to incorporate the code into their own work is an additional requirement for openness. Some suggest that sharing code but not documenting it or making obvious how it works does little to make the code truly open. As this book shows, debates over the meaning of *openness* have been a key catalyst for developments in FOSS history. In this respect, the history of FOSS reflects a wider debate about the meaning and importance of openness in the computing world since the 1970s. Scholars such as Andrew Russell have explored this subject generally but without extensive focus on the FOSS context.[3]

Notwithstanding the complexities of basic FOSS terminology, this book aims to use the terms *free software* and *open source* in ways that are consistent with the software creators' intentions. In these pages, the term *free software* refers to software whose programmers or users call it such because its source code can be studied and modified freely by people who use the software, whether or not the source code costs money. Some proponents of free software write the term with a capital *F* as *Free software* to emphasize that the software is "Free as in Freedom" and not necessarily free of cost.[4] For the sake of stylistic consistency and political objectivity, however, this book does not capitalize the term.

Meanwhile, I use the term *open source* in reference to software whose creators and users preferred that term over *free software*. I do so whether or not the software in question or the license that governs it is actually different in a qualitative way from what constitutes free software. In other words, I have tried to adopt the usage of the historical actors themselves whenever possible.

In this book, the term *free software* does not mean software that is available without cost but that shares no affinity with what early proponents of the free software movement called *free software*. Software that is simply given away for free is best labeled *freeware*. In contrast, free software makes it possible for people other than the original creators of the code to share, study, and modify the software. For these purposes, access to source code is a prerequisite, because source code is necessary for understanding how a program works and for making changes to it. In addition, although FOSS usually costs no money, being free of cost is not a defining characteristic of either free or open source software. As this book shows, some of the most prominent FOSS projects have not always shared their products free of cost. In other cases, payment has been optional. It is the public availability of source code, not pricing, that defines FOSS.

## THE HISTORY OF FOSS HISTORY

Few books are as simultaneously easy and difficult to write as one about the history of FOSS. On the one hand, because no researcher has yet attempted to narrate, in a systematic, comprehensive, and objective way, the story of the FOSS movement from its origins amid post–World War II "hacker culture" through the present, this book digs into fertile ground. That is a rarity because most works of history—even those dealing with recent time periods, as this one does—tend to engage topics that have already received so much attention from researchers that the best historians can usually hope to do is turn over a new stone or reinterpret an existing argument, not present a great deal of original material. That this book seeks to do the latter has made it a privilege and a pleasure to write.

On the other hand, the lack of any published history of the FOSS movement that aims to be both comprehensive and objective in its perspective has also made this a difficult book to research and write. A significant body of work that engages FOSS history exists, but none of it corresponds directly to the goals of this book. To make clear what this book does differently, it is worth briefly surveying previous studies that deal with similar topics.

### FOSS Primary-Source Histories

Existing published work on the history of FOSS comprises two main bodies of literature. The first consists primarily of primary sources produced by leading figures from FOSS history or by journalists and researchers who worked closely with them. These works include Eric S. Raymond's essays "A Brief History of Hackerdom" and "The Revenge of the Hackers," which tell the story of FOSS from their author's perspective between the late 1960s and the end of the 1990s.[5] Also notable are biographies of Richard Stallman and Linus Torvalds, which narrate the contributions of those luminary figures to the development of FOSS.[6] Peter Salus's 1994 work *A Quarter Century of UNIX*, which contains lengthy excerpts of interviews with Unix programmers, offers useful primary material, along with informative commentary by Salus, on the history of certain Unix-like operating systems (although Salus barely mentions GNU or Linux).[7] The essays published in Chris Dibona, Sam Ockman, and Mark Stone's edited 1999 volume *Open Sources: Voices from the Open Source Revolution*, as well as Glyn Moody's 2002 book *Rebel Code: Linux and the Open Source Revolution*, which also contains extensive interviews with FOSS programmers, round out the list of major published primary-source accounts of FOSS history.

Most of these primary-source works are subject to two chief limitations. The first is that they were written by or in close collaboration with the figures they describe. That does not necessarily mean the stories they present are inaccurate or their claims are biased. But it does constrain their perspective, and it raises flags for any historian conscious of the dangers of blindly taking historical actors at their word. By critically analyzing the essays and biographies of figures like Raymond, Stallman, and Torvalds and pointing out the contradictions between those works, this book endeavors to tell the story of FOSS from a broader, more objective point of view than that of earlier attempts.

In addition, this book draws not only on the well-known primary texts described above but also on more obscure primary sources, such as Usenet posts, email archives, and original interviews that I conducted with FOSS programmers. It therefore brings to light a wealth of evidence that presents a deeper, more nuanced explanation of the history of FOSS than figures such as Raymond and Torvalds have been able to provide in their memoirs.

The second limitation of the existing primary-source histories is that virtually all of them were written in the early 2000s or before. That is unsurprising. The late 1990s—when the term *open source* became a common part of the tech lexicon and companies selling Linux or related software enjoyed explosive growth—remain perhaps the most exciting and promising period in the history of FOSS. Developments since that time— such as Canonical's attempt to reinvigorate desktop Linux by introducing Ubuntu in 2004 and the behind-the-scenes role that Linux has played in Android smartphones and tablets— have not generated as much public attention to FOSS. Yet that does not mean that what has happened in the FOSS world since the early 2000s is not important for explaining how FOSS developers, companies, and supporters arrived where they are today. Accordingly, this book aims to weave together the disparate histories of FOSS that were produced before the early 2000s and also extend the story they tell into the present.

### Scholarly Works

Scholarly works represent the second main body of published literature about FOSS history on which this book draws. Beginning in the early 2000s, academic researchers explored some dimensions of FOSS history as part of studies that were primarily sociological or anthropological in focus.

The first overview of FOSS history intended for academics appeared in 2001, when David Bretthauer, a librarian at the University of Connecticut (and a member of the Connecticut Free Un*x Group), published "Open Source Software: A History."[8] This brief essay mostly rehashed information from the published primary sources discussed above. Yet it was significant because, unlike earlier works, it attempted to provide a complete narrative of FOSS history. It described the cultural and philosophical origins of FOSS values, as well as the development of GNU, Linux, and the BSD (Berkeley Software Distribution) operating system. Bretthauer also briefly touched on the histories of some other important FOSS projects, such as Perl, Python, and Apache. His essay offered few details, but it encouraged scholars for the first time to think of FOSS as a singular phenomenon with a long history worth exploring.

Working in the same vein, Nathan Ensmenger became the first scholar to discuss the academic significance of FOSS history in an explicit way. In a 2004 article titled "Open Source's Lessons for Historians," Ensmenger pointed out how FOSS communities provide "new ways for ethnographers and political scientists to think about the process of self-governance."[9] He also suggested studying FOSS as a means of understanding how "technical, social, and political agendas" have been deeply intertwined with software development since the early days of computing.[10] Most significant was Ensmenger's contention—despite the oversimplified view of FOSS history that prevailed in 2004 and remains largely intact today—that the advent of FOSS did not represent a sudden, "radical break" with older ways of writing and distributing code. He argued that closed source software and FOSS share a longer, intertwined history. They are not dichotomous opposites, he wrote, but different stretches along the "continuum of possible configurations of technology, markets, and intellectual property regimes."[11] This book expands on that idea by highlighting the diverse approaches to software development and distribution that have emerged within the FOSS world, some of which share more in common than others with proprietary software.

Ensmenger's article presented insights that are crucial for understanding where FOSS fits within the complex history of computing and software. It also challenged the oversimplified narratives of FOSS's past that writers such as Raymond presented. Yet Ensmenger's three-page essay was a "think piece" and did not analyze the ideas it introduced in any detail. Ensmenger subsequently revisited some of the article's theories in his excellent 2010 book, *The Computer Boys Take Over*. That work, however, focused not on the origins of FOSS but on the history of the professionalization of programming.[12]

Steven Weber did more to examine the ethnographical and political theories of FOSS developers in *The Success of Open Source*, the first book-length scholarly study of FOSS. Weber's chief interest was exploring how property "underpins the social organization of cooperation and production in a digital era."[13] To perform that study, he leveraged FOSS as "a real-world, researchable example of a community and a knowledge production process that has been fundamentally changed, or created in significant ways, by Internet technology."[14] In the course of investigating how the Internet has shaped the way FOSS developers produce code and think about property, Weber recounted the histories of Unix, BSD, and Linux from their origins through the early 2000s.

Because Weber's goal was not to provide a complete history of FOSS, his book paid considerably less attention to the history of other FOSS projects, such as GNU. In addition, as the title of Weber's book suggests, he was concerned mostly with FOSS projects that were successful. He did not analyze the limitations of FOSS development or provide detailed explanations for why some FOSS projects succeed while others do not. In that respect, this book offers a corollary to Weber's text. For example, one of the major tasks of this book is to explain why GNU development lagged in the 1990s while Linux surged.

Christopher Kelty's *Two Bits: The Cultural Significance of Free Software* pursued an agenda similar to Weber's.[15] Published in 2008, the book represents the second major scholarly analysis of FOSS. Kelty worked from an anthropological perspective. His chief argument is that FOSS communities function as "recursive publics" that are "concerned with the ability to build, control, modify, and maintain the infrastructure that allows them to come into being in the first place."[16] Like Weber, Kelty engaged FOSS history only to the extent necessary to facilitate an anthropological analysis rather than to explain it comprehensively. Still, he narrated the histories of Unix, GNU, and Linux relatively thoroughly, and his work is of vital relevance for placing historical FOSS developments within cultural and social contexts.

Several other scholarly works that do not engage extensively with the history of FOSS nonetheless offer valuable perspectives for this book. They include Russell's *Open Standards and the Digital Age*, which, as noted above, explores how "openness," broadly conceived, has affected engineers over the past several decades.[17] Martin Campbell-Kelly's history of software between the 1950s and 1995, *From Airline Reservations to Sonic the Hedgehog*, offers excellent perspective on the crucial effects that software in general, if not FOSS in particular, have exerted on society since the early days of computing. Mark Priestley's *A Science of Operations: Machines, Logic and the Invention of Programming* is valuable for similar reasons. An article by Charles Yood, which argues that "the history of computing provides a rare opportunity to make the work of historians relevant and interesting to people outside the academy," helped to inspire this book, which was written in the hope that academics and nonacademics alike will find value in its pages.[18]

Lastly, works by David Berry, Wendy Chun, Federica Frabetti, Matthew Fuller, Eric von Hippel, Lev Manovich, and Georg von Krogh on the cultural, economic, and philosophical significance of software are vital for evaluating the roles that FOSS programmers and users have played over the last several decades within broader ecosystems of software, computers, and code.[19] This scholarship helps to explain why FOSS has meant so much to so many people. As these scholars have noted, software's significance extends far beyond the mathematical operations of code itself. Manovich argues that software has acquired an omnipotent relevance by becoming the basis for all forms of media in the digital age.[20] Chun notes how access to and control over code seem to be the only means of managing the endlessly complex and obscure workings of the computers that run our lives—even if, in reality, the relationship between code and a computer's actions is not at all straightforward because code can be translated by compilers or executed by machines in unexpected ways.[21]

Against this backdrop, the stakes of the debates about FOSS, which this book narrates, come into clearer focus. On their own, the utilitarian dimensions of code do not explain why FOSS programmers and users have so passionately debated one another and their counterparts in the closed source ecosystem. Being able to borrow and reuse code, as FOSS developers do, is a convenience. It saves time and is arguably more efficient. But it is not important enough on its own to motivate rational people to stake personal reputations and fortunes on FOSS code when closed source software can be just as profitable. Programmers could still program and computer users could still run their code, even if all code were closed.

What they would lack under those circumstances is the sense of exerting control over their world. By retaining full access to the source code of the software that runs much of their lives, FOSS developers and users seek independence. Even if they never actually modify programs, knowing that they can do so is tremendously assuring. That is the principle reason that so many people have cared deeply about FOSS for several decades. The works cited above do not study how this thinking has affected the FOSS community in particular, but they provide the foundation for interpreting the FOSS approach to software in this vein.

### SCOPE AND THEMES

Although this book is the most comprehensive study to date of the history of FOSS, its scope is necessarily limited in several ways. For one, it does not discuss the individual history of every FOSS project. That would be impossible. A 2012 analysis found that the number of FOSS projects in existence totaled between 324,000 and 4.8 million, depending on how a distinct project was defined.[22] The figures have grown since then.

Nor does this book afford equal treatment to all theaters of FOSS development. It offers relatively little discussion of FOSS programs that are designed to run on Microsoft Windows, for example. Instead, the text focuses on those projects and communities—notably Unix, BSD, GNU, Linux, and Apache—that have been most influential in shaping the technical, intellectual, and cultural dimensions of FOSS.

It touches along the way on the histories of many other projects, from Perl and Python to Wine and X Windows, that will also be familiar to many FOSS users. Unfortunately, however, for want of infinite space, this book skips over some FOSS programs that are historically significant. I hope that the future will present opportunities for historians to study these projects at greater length.

Narrating the major developments of FOSS history over the past half century is only one part of this book's focus. It seeks also to explain why events occurred as they did—why a certain group of programmers in the 1980s decided to give away potentially valuable code free of cost, why the Linux kernel evolved from obscure origins to become one of the most widely used software programs in the world, how the FOSS community overcame attempts by companies like Microsoft to discredit it in the late 1990s, and much more. Interpreting and accounting for major developments in FOSS history such as these is crucial because to date few observers have considered why FOSS followed the path it did.

Last but equally significant, the book reconsiders some of the stereotypes that have affected the FOSS community. One of those is the relationship that FOSS developers have had with commercial endeavors. It has been common practice to label FOSS advocates as anticapitalists who choose to share their code freely primarily because they disdain economic exchange.[23] Yet the reality is much more complex. Historically, some free software purists, including those associated with the Free Software Foundation, collaborated enthusiastically with companies and entrepreneurs to find ways to make money with software that they gave away for free. In other cases, such as that of the Linux kernel early in its history, FOSS leaders who became famous for ostensibly pragmatic attitudes toward the commercial use of FOSS proved to be reluctant to engage in any activity that smacked of commercialism. In the latter instances, having "fun," a word that FOSS developers and users have frequently used to explain their interest in freely shared code, was a more important impetus for participation in FOSS projects than was making money.

It is from this complex interplay between pleasure and commercial interest that this book takes its title. The history of FOSS is the history of programmers and users who were motivated by both fun and profit, not one or the other, as conventional interpretations of the FOSS ecosystem have tended to assume.

In a similar vein, this book illuminates the fascinating ways in which the FOSS community has navigated the straits separating pragmatism from ideology over the last several decades. Some observers have been quick to dismiss certain FOSS activists, such as those associated with the Free Software Foundation, as ideologues incapable of compromise or cooperation with people who espouse alternative viewpoints. There also has been a tendency to contrast such groups with the supposed pragmatists in the "open source" camp, whose willingness to compromise ostensibly ensured their success. This book shows that such depictions miss the mark. To be sure, FOSS proponents have not always acted pragmatically. Their obstinacy has sometimes been detrimental to their endeavors. Yet an understanding of the interplay between pragmatism and ideology within FOSS history requires an appreciation of the extensive nuances that have marked the evolution of FOSS projects and culture over the past several decades. It is not true that one part of the FOSS community has always been overly idealistic while the other has consistently exercised cool pragmatism.

## FOSS AS A REVOLUTION

As the first paragraphs of this introduction suggest, this book uses revolution as an interpretive lens for evaluating the history of FOSS. This approach reflects two main themes that run consistently throughout this book.

The first is that FOSS leaders and community members have frequently described themselves as revolutionaries and deployed the rhetoric of revolution to explain or justify their actions. Torvalds and Raymond both have referred to themselves as "accidental revolutionaries."[24] Similarly, Raymond borrowed a phrase from poet Ralph Waldo Emerson about the start of the American Revolution when he described the events of January 1998, which jump-started the open source movement, as the "shot heard 'round the world of the open-source revolution."[25] A

2001 documentary called GNU/Linux the revolution OS (operating system).[26] The first major anthology of writings about open source history from the leading figures of the movement characterized the events as "the open source revolution."[27] Without appreciating how important the concept of revolution has been to FOSS leaders and supporters, this book could not accurately interpret their decisions and ideologies.[28]

Revolution also serves as an effective interpretive lens for understanding the history of FOSS because so many of the events and trends in FOSS history have followed a "revolutionary script" that was similar to that followed by the major political revolutions of modern times.[29] Although changes to the ways that software is written and distributed constitute a bloodless revolution, unlike many political revolutions in which social hierarchies and cultures are violently contested, the story of FOSS closely parallels other revolutions of recent centuries.

For example, the FOSS revolution's origins can be traced to marginal figures who seethed with resentment against mainstream modes of software distribution in the 1970s and early 1980s. They believed that artificial access barriers to source code stifle creativity and sustain arbitrary hierarchies. In this sense, people like Stallman, whose name was little known within the programming world before the launch of GNU, were not unlike French revolutionaries such as Jean-Paul Marat or Jacques-Pierre Brissot. As the historian Robert Darnton has shown, the hatred toward the old regime that Marat and Brissot exhibited prior to the French Revolution, when they were struggling, marginalized writers, was born in part from their perception that aristocratic society prevented them from participating in intellectual life in the way they believed they deserved.[30]

Similarly, the conflicts that have shaped the history of FOSS parallel the wars that accompanied most major political revolutions. The American Revolution centered on a violent struggle between American rebels and military forces from overseas who were loyal to the British crown, which was complicated by battles between the colonial rebels and Loyalists. The French Revolution witnessed brutal, multifactional civil war between royalists, federalists, and Jacobins, which coincided with the existential struggle that the French revolutionaries waged against foreign armies. The 1917 revolution in Russia began in the throes of a devastating war against foreign powers and then was transformed into a prolonged domestic struggle that pitted Bolsheviks against White Russians while foreign powers meddled from the sidelines. War against domestic and foreign enemies proved inseparable from each of these political revolutions.

In the same way, as this book shows, the FOSS revolution included two major "wars." One was waged by FOSS supporters against Microsoft and associated companies, which sought in the late 1990s and early 2000s to stamp out the FOSS movement completely. In the same period, a prolonged struggle within the FOSS community proceeded as supporters of the Free Software Foundation vied with those in the "open source" camp to define the meaning and scope of FOSS development. Comparing these conflicts with the role of warfare in political revolutions provides useful perspective for understanding how and why different types of confrontation have shaped FOSS.

The outcomes of major political revolutions also offer perspectives on that of the FOSS revolution. In many cases, the people and ideologies that revolutions ultimately usher into power are very different from those that prevail at the beginning of a revolution. The liberal aspirations of the constitutional monarchists who launched the French Revolution in 1789 gave way to the radical bloodshed of the Reign of Terror, followed eventually by Napoleonic dictatorship. The Russian Revolution descended from a movement focused on moderate change and reform and evolved into one that radically restructured society.

The FOSS revolution's trajectory has similarly involved a number of revolutionary "stages" and shifts in the center of power. The FOSS movement began in the early 1980s under the leadership of Stallman and other nonconformist hackers at the Massachusetts Institute of Technology in a loose alliance with like-minded counterparts at the University of California at Berkeley. Their methods and values contrasted in certain key ways with those that emerged starting in the early 1990s within the Linux development community, out of which the open source (as opposed to free software) movement emerged in the later 1990s. The open source faction's commercial success paved the way for FOSS projects like Linux and Mozilla Firefox to become household names. In contrast, Stallman's GNU operating system project—despite continuing to provide a great deal of the code that makes FOSS platforms run—is much less well known among the public at large today. The bulk of FOSS development and investment now centers on groups whose members do not necessarily espouse a great deal of interest in the values that Stallman articulated when he launched the GNU project and the Free Software Foundation in the 1980s. In this way, the people and ideas that launched the FOSS revolution have not retained definitive control over it.

## ORGANIZATION

This book begins with a study of the origins of hacker culture and its influence on the Unix operating system, which are the subject of chapter 1. Although Unix at the time of its creation was never called free or open source software, it effectively functioned as such in key respects. It was developed collaboratively in disparate locations by programmers who shared the code openly with one another. Hacker culture, which had its roots in a period somewhat earlier than Unix's creation and shared affinities with the academic culture from which it was born, found a strong footing within the Unix community during the late 1960s and 1970s. The community of Unix hackers thrived until the early 1980s, when AT&T, the company that owned Unix, turned the operating system into a commercial product, engendering a crisis among hackers.

Chapter 2 describes the reactions to that crisis. One of them involved an effort at the University of California at Berkeley to write a clone of Unix called BSD (Berkeley Software Distribution) that was free of AT&T code, allowing it to be distributed in ways that hackers could accept. (As the chapter explains, BSD development initially focused on enhancing rather than replacing Unix, but Unix's commercialization transformed the mission of BSD programmers.) Another reaction, which fomented the birth of the free software movement as a conscious initiative, was Richard Stallman's endeavor to build a separate Unix clone, which he called the GNU operating system. The focus of chapter 2 is the GNU project's activities throughout the 1980s and related developments, such as the founding of the Free Software Foundation and the establishment of free software licenses.

Chapter 3 considers the events of the early 1990s. Although most hackers at the time believed that either BSD or the GNU operating system was poised to provide a platform that would be an alternative to Unix and would provide source code freely for a reasonable price, unexpected turns stunted both of these projects. BSD came under legal fire, which scared some potential contributors and users away. Meanwhile, although GNU developers were highly successful in writing many other programs, they lagged in their efforts to create a kernel, the core part of an operating system. As a result, the opportunity arose for Linus Torvalds, a young programmer from Finland, to produce a kernel of his own that rapidly became much more popular than its obscure, hobbyist origins suggested it should have. Chapter 3 narrates the history of early Linux development and considers why Torvalds and the community of programmers he led assumed such outsize influence within the FOSS community in the early 1990s.

Chapter 4 discusses what happened when the suite of software programs produced by the GNU project combined with Linux during the 1990s. Together, GNU software and the Linux kernel fueled the emergence of a booming new sector within the technology industry. Companies invested billions of dollars in business operations that centered on software that was given away for free. This activity cemented the position of FOSS within the mainstream technology world and made FOSS operating systems a viable choice for millions of people.

Yet as chapter 5 shows, the growing popularity of FOSS in the 1990s also stoked deep divisions. Tensions arose within the FOSS community regarding what the values of FOSS developers and users should be and what constituted FOSS code. These divisions also led to a battle between FOSS supporters and closed source software companies, especially Microsoft, which had come to see FOSS as a threat to its business. Chapter 5 discusses both of these "revolutionary wars."

The final chapter explores developments that have occurred in the FOSS world since the revolutionary wars of the late 1990s and early 2000s ended. Following the calming of internal political tensions and the disappearance of external threats, the FOSS community has enjoyed remarkable momentum over the last fifteen years. This has been true not only in the niches, such as Internet servers, where FOSS established commercial dominance early on, but also in new areas, including embedded computing, mobile devices, and the cloud. In addition, as chapter 6 also explains, FOSS has left a profound cultural and intellectual mark on other initiatives that have no direct relation to the software industry, such as *Wikipedia* and Creative Commons. Yet supporters of FOSS continue to debate the goals of the FOSS revolution and whether they have been achieved. In addition, technology has evolved in ways that previous generations could barely have foreseen, necessitating new strategies for FOSS developers. As a result of the lack of consensus about FOSS and the shifting technological landscape, the FOSS revolution continues. So does the struggle of women and minority programmers to enter FOSS communities, a topic also discussed in this chapter.

# 1 THE PATH TO REVOLUTION
## Unix and the Origins of Hacker Culture

MOST ASPIRING REVOLUTIONARIES thrive on visions of a mythical golden age of the past, which they deem it their duty to resurrect. Colonists during the American Revolution hoped to restore the right to representation that they thought they merited as British subjects—even though Great Britain has never had a formal constitution that guarantees such rights. Two decades later, French revolutionaries saw their mission as recovering the natural laws that, according to ancient writers, had prevailed in a mythical time when all men were equal and lived in harmony.[1]

Resurrecting a lost golden age has been a significant part of the way that FOSS proponents have thought about their work. Activists from Stallman to Raymond believed that, by advocating for open and freely shared code, they were working to restore the moral principles that predominated in the early days of computing—when the values of self-described hackers, rather than business interests, defined the way programmers produced and distributed software.

Understanding the ways in which the early FOSS community perceived this golden age of software sharing involves examining two intertwined segments of the history of computing.

The first centers on Unix, the operating system that emerged in the late 1960s and nurtured many of the values and practices that greatly influenced the development of FOSS software in later decades. The second involves the origins of what Raymond and other writers have described as the "hacker ethic," which was born long before distinctions arose between free and closed source software—and which, as I argue below, owed much more to the influence of academia than hackers have tended to recognize. This chapter details both of these topics and describes how they set the stage for the FOSS revolution of subsequent years.

## UNIX AND FOSS

It might seem strange to begin a history of FOSS by discussing the early years of Unix. Even when Unix could be freely shared, it was not described as *free* or *open source software*. These terms did not yet exist. Moreover, Unix ended up becoming the opposite of FOSS after it was transformed from a research project into a commercial product in the mid-1980s. None of the pioneers of Unix development went on to become major figures in the FOSS movement. And Unix code formed no significant part of GNU software, the Linux kernel, or other FOSS platforms. On the contrary, the FOSS revolution began because the programmers involved in these later projects sought to build operating systems free of Unix code.

Nonetheless, for the first fifteen years or so of Unix's existence, it bore all of the hallmarks of the major FOSS projects that followed it. It was effectively free and open source software before such terminology came into use because Unix's code was collaboratively developed and freely shared among programmers at a number of different locations. The Unix community adopted this practice because it was the norm at the time of Unix's birth. Unix arose before the commercialization of software prompted anyone to conceptualize a difference between free and nonfree software or to suggest that sharing code openly was not the natural thing for programmers to do.

Unix was also similar to many of the major FOSS projects, especially GNU, in that it was the brainchild of somewhat disillusioned programmers who, working on their own initiative, sought to fill a void that affected them personally—in other words, "to scratch a personal itch," to borrow the parlance that Raymond introduced in the 1990s to explain what drives open source developers.[2] Unix's founders wanted an operating system that offered all the best features of the failed Multics platform (for more on Multics, see below) and ran on whatever computer hardware they had on hand in their office.

Finally, like Linux and so many other FOSS projects, Unix grew out of a coding effort that was undertaken "just for fun," in the words of one of its founders. This was the same phrase Torvalds used to describe his approach to programming many years later.[3]

For all of these reasons, understanding the role that Unix played during the late 1960s and 1970s in giving rise to both the technical and cultural ideas that later became central to FOSS development in the 1980s and 1990s is crucial for foregrounding the emergence of FOSS projects properly defined.

Before delving into the details of Unix's birth, however, it is worth clarifying what, exactly, the word *Unix* signifies for the purposes of this book. The term has acquired an ambiguous meaning in the decades since the operating system's appearance, mainly because programmers and authors sometimes use *Unix* as a shorthand to refer to operating systems that were designed to function like Unix and may or may not derive in part from the original Unix code base. For example, the Berkeley Software Distribution (BSD), which initially shared much of its code with Unix (later, BSD developers replaced all of the Unix code with their own, original software) is sometimes labeled *Unix*. So is Linux, which never shared any code with Unix but has central features that were modeled on those of Unix.

In most cases, operating systems that resemble Unix but are distinct from it should be called *Unix-like systems*. That is the usage this book adopts. Throughout these pages, the word *Unix* refers to the operating system that was built at AT&T's Bell Laboratories beginning in 1969 and released by AT&T in a number of different versions until 1990, when AT&T spun off its Unix division into Unix Systems Labs.[4]

## IT CAME FROM OUTER SPACE: THE ORIGINS OF UNIX

Programmers began developing Unix because they wanted to play a game called Space Travel. Unrelated to the better-known Spacewar game that some writers have erroneously associated with Unix, Space Travel simulated voyages within Earth's solar system.[5] In 1969, Ken Thompson, a programmer at Bell Laboratories (popularly known, then and now, as Bell Labs) in Murray Hill, New Jersey, was writing Space Travel to entertain himself. Thompson was twenty-six years old at the time and only three years out of the University of California at Berkeley, from which he had received bachelor's and master's degrees in computer science in 1965 and 1966, respectively.[6]

After completing his education, Thompson had been part of the Bell Labs development team for the Multics operating system, on which Bell had collaborated with General Electric and Project MAC, a leading computer science research facility at the Massachusetts Institute of Technology. Launched in 1964, Multics (shorthand for Multiplexed Information and Computer Services) was an ambitious project that aimed to build a next-generation time-sharing operating system. Time-sharing systems, which allow multiple users to log in to a single computer at once, were a relatively novel idea in the mid-1960s. Multics was not the first time-shared system, but its developers hoped it would provide more reliable and convenient computing than the existing alternatives. They also sought to pioneer a simpler and more streamlined way for computers to handle storage objects, manage processes, and communicate with disk drives and other peripheral devices.[7]

Multics development continued through the mid-1980s, and the platform was eventually deployed in production environments. But it failed to achieve the momentum and level of innovation that its designers had intended. Frustrated with the lack of progress, Bell Labs withdrew from the project in 1969. The move demoralized Thompson and other programmers who had poured their talents into Multics coding.[8]

Yet like so many FOSS developers who followed in their footsteps a generation or two later, the Bell Labs programmers did not allow the decision of corporate managers to sever ties with the Multics project to dash their hopes of completing work on the innovative technical features they had begun implementing. Spurred on by their desire to run Thompson's Space Travel game on a Digital Equipment Corporation PDP-7 machine—a comparatively inexpensive computer that Bell Labs had assigned to them after the collapse of the Multics development effort on more sophisticated hardware—Thompson and Dennis Ritchie, another member of the former Multics team, began writing a very basic operating system for the PDP-7 that could host the game.

Working in assembly language, which was tedious but provided the programmers fine-grained control over the way the computer executed code, Thompson and Ritchie developed a file system for the PDP-7 that mimicked the one intended for Multics. They accompanied it with simple utilities for copying, printing, deleting, and editing files. They created a shell program, which provides an interface through which users can run commands and view output. They integrated system processes into the system, which allowed multiple programs to run at the same time—another major innovation originally planned for Multics.[9]

In its earliest incarnation, the operating system that Thompson and Ritchie wrote for the PDP-7 was as basic as basic could be. Designed merely to run the Space Travel game that Thompson had written in his spare time, the platform had little obvious research applicability or commercial promise. It did not even have a name until 1970, when Bell Labs programmer Peter Neumann suggested *UNICS* to characterize the system as an emasculated alternative to Multics. (Later in the same year, fellow Bell developer Brian Kernighan suggested *Unix*, the version of the name that stuck.)[10] And because the system was written in assembly language, which depended heavily on features specific to the PDP-7 hardware, it was hardly suitable for porting, or adapting software designed for one type of computer to run on a different kind. (Linux in its first incarnation was similarly intended never to be portable, adding to the irony that both Unix and Linux eventually became some of the most portable operating systems of their times.)

Yet even without a name, a clear future, or anything approaching advanced features, the amalgam of basic storage tools, system processes, and shell utilities that Thompson and Ritchie had hacked together in creating Unix represented something novel. By breathing life into some of the most innovative features of Multics, these developers had built a platform that promised to interest programmers and systems administrators at places far beyond Bell Labs. They had birthed a new type of system—along with the beginnings of a special culture to accompany it—that would evolve into something much more important than a way to play a space voyage game.[11]

### UNIX GROWS

Unix evolved rapidly following its creation at Bell Labs in 1969, and so did the culture that grew up around it. This was the environment in which many of the paradigms and philosophies that defined the FOSS projects of succeeding decades had their first direct incarnations.

In 1970, after Bell Labs purchased a newer PDP-11/20 computer, Thompson and Ritchie ported Unix to run on it. Working with that updated hardware, they implemented more utilities, including enhanced text formatting and editing programs. To help other developers use the platform, they released the first Unix programmer's manual in November 1971. It introduced the manual ("man") page formatting that GNU/Linux and other Unix-like operating systems continue to use today.[12] By 1973, Thompson, Ritchie, and other collaborators had rewritten most of the Unix code in C, a programming language that Ritchie had begun developing the previous year with the goal of making Unix easier to port to other platforms.[13]

Although Bell Labs remained the center of Unix development during the operating system's early years, programmers at other sites across the world quickly took an interest in the project and began contributing code and ideas back to the platform's main developers. Collaboration became easier in 1973, when AT&T licensed Unix for use by educational institutions, where many programmers were employed. By the end of the year, the operating system had spread to universities across the United States and to at least eleven institutions in the United Kingdom. The University of New South Wales, Australia, became the first site in the Eastern Hemisphere to adopt Unix, in 1974.[14] Commercial organizations began using the operating system in 1975.

Bell Labs rolled out new versions of Unix rapidly throughout this period, establishing the "release early, release often" paradigm that became a distinguishing feature for many FOSS projects in later decades, as Raymond noted (although without tracing the innovation to Unix development) in his seminal essay on open source development practices, "The Cathedral and the Bazaar."[15] In fact, Thompson declared that, from the perspective of Bell Labs, Unix development "was a continuum," with developers constantly introducing enhancements to the operating system, rather than a series of distinct releases, even though the system was distributed to other sites in specific versions.[16]

That characteristic, which anticipated the "rolling release" model that many FOSS projects follow today (as well as, to a certain extent, the "continuous delivery" paradigm of the modern DevOps movement), set Unix apart from most major software projects of its time. Other developers tended to release new versions of their software slowly, after tedious and time-consuming testing and debugging periods. In contrast, because most of the organizations using Unix participated actively in bug-reporting and development processes, Unix developers were able to prioritize quick releases and rapid innovation over a vain quest for perfect stability, which has bogged down many other software development efforts, then and now.

The pace of innovation in the Unix community increased further in early 1978, when Bill Joy, then a graduate student at Berkeley and later an executive at Sun Microsystems, started work on what grew into the Berkeley Software Distribution (BSD). The system began as a variant of Unix that mixed code from Bell Labs with programs written at Berkeley to enhance Unix's functionality. In March 1978, Joy began distributing BSD to sites beyond Berkeley, providing them with an alternative to AT&T's Unix.[17]

Berkeley's emergence as a second center of Unix development—and one that, until the 1980s, focused on producing code that complemented the features present in the Bell Labs version of the operating system rather than replicating them—solidified Unix's position as an innovative operating system that computer scientists at sites across the world were collaboratively developing. Each local programmer or team of programmers added improvements to Unix that made the system a better fit for the particular hardware or software environments at their sites. In turn, they shared their changes with Unix developers and users at other locations, ensuring that the best innovations found their way back into the main code base. That approach to development—which depends centrally on sharing code publicly and freely, unencumbered by restrictive licenses or patents—was much the same as the one that fuels most FOSS projects today.

The culture that the Unix community fostered also prefigured the one that sustains the FOSS world today. By the late 1970s, Unix user groups appeared, organizing meetings and activities similar to those of GNU/Linux user groups today.[18] (After Bell Labs complained in 1977 that Unix user groups did not have permission to use the *UNIX* trademark in their names, the community adopted the term *USENIX* as a playful replacement.) In addition, Unix users began circulating a newsletter, originally called *UNIX NEWS* and renamed (also as a result of trademark issues) *;login* in 1977.

Also significant was that the Unix community conceptualized itself as a rebel force. In the view of Unix advocates, the operating system constituted a bulwark against the encroachment of business interests on the computer world. In this vein, Armando Stettner, who participated in Unix development both at Bell Labs and as an employee of Digital Equipment Corporation (the company that manufactured many of the computers that were used by Unix developers in the 1970s and 1980s), distributed license plates that channeled those of the state of New Hampshire by declaring "Live Free or Die: Unix."[19] As Raymond recalled, Unix developers "liked to see themselves as rebels against soulless corporate empires."[20]

Such sentiments deepened after AT&T began selling Unix as a commercial product in the early 1980s (a change discussed at greater length below). The commercialization prompted free-minded Unix developers to rally around the BSD variant and begin turning it into a standalone Unix-like operating system that did not rely on any of AT&T's code. *Star Wars*–themed posters depicted BSD developers as holdouts against imperial commercial interests. Like some FOSS hackers today, the anti-corporate Unix enthusiasts hoped to provide the world with software that was free to be the best it could be technically rather than software that business managers deemed to be the most profitable or the easiest to sell to a mass market.

The Unix community's resistance against what its members perceived as arbitrary restrictions on the way they could use software translated into more than posters and license plates. Its members also engaged in activities that directly subverted the barriers imposed by proprietary software. For example, in 1977, Tom Ferrin, a Unix programmer based at the University of California at San Francisco, published an article in the *;login* community newsletter that described a hardware modifica-tion that Unix systems administrators could perform on PDP computers by cutting a piece of foil and inserting a jumper wire. The modification resolved a bug that occurred on these computers as a result of "DEC's desire to 'preserve the integrity of proprietary programs'" rather than allowing third parties to modify them, according to Ferrin.[21] Such efforts anticipated practices that are common in the FOSS community today, where developers evince few qualms about independently reverse-engineering device drivers or software protocols in order to use their computers as they see fit, even if the original designers of the software or hardware do not condone their activities.[22]

## UNFREE UNIX

At first, the Unix community's cohesiveness stemmed largely from the quirkiness of the development model that Unix developers embraced, as well as the core design philosophy of the operating system—which emphasized modularity and the principle that each part of the system should focus on doing a specific job and doing it well. At the time, that approach represented a novel way of thinking about operating system design. It helped to bind together the community of Unix programmers and users. Meanwhile, the fact that AT&T's policy was to distribute Unix with "no advertising, no support, no bug fixes [and] payment in advance" meant that members of the Unix community had to rely on each other for help in deploying the software. That dependence also fostered greater cohesion and a sense of being different from other parts of the computing world.[23]

Yet the rebellious spirit of the Unix community assumed new urgency in the early 1980s, when the licensing terms surrounding the operating system changed dramatically. When Unix was born in 1969, Bell Labs, as a subsidiary of AT&T and Western Electric, was subject to a consent decree that the United States government had issued on January 25, 1956. The decree forbade the two companies and their subsidiaries from engaging in business unrelated to telephone or telegraph service or equipment. It permitted other activities only if they were not tied to commercial operations. Because Unix had nothing to do with the telephone or telegraph business, Bell Labs could legally develop it only if it remained a research project rather than a commercial endeavor.

This was why, until the mid-1980s, AT&T allowed the Unix programmers at Bell Labs to distribute their code to universities and later to commercial organizations for a nominal fee. It was also why Bell Labs could not offer support services for Unix, as these would have been unrelated to AT&T's telephone and telegraph business. The lack of support meant that external organizations that used Unix took a keener interest in having access to the code and in enhancing the system to meet challenges that they faced at their particular sites than they likely would have if they could have contracted with AT&T to install and configure the software for them. In this way, the consent decree ensured that Unix would remain a research project and also that Unix development would depend on a decentralized community of global collaborators.[24]

All of this changed in 1984. In that year, as the result of an antitrust action against Western Electric and Bell Labs that the United States government had initiated in 1974, the companies that owned Bell Labs were reorganized, and the consent decree of 1956 no longer applied.[25] The new policy prompted AT&T to revise the Unix licensing terms and promote System V, the version of Unix it was developing at the time, as a commercial product. By 1988, the price for a System V source license, which provided access to the system's full source code, exceeded $100,000, five times what it had been a few years earlier. It had grown to nearly double that amount five years later.[26] Commercialization changed the operating system irrevocably.

The Unix story did not end in 1984. Unix development continued well into the next decade, even as ownership of the code passed from AT&T to a succession of other companies and organizations. For the purposes of this book, however, there is not much to say about Unix after it became a commercial product. Following that shift, Unix's chief importance for the FOSS community was as the embodiment of something to fight against or to try to outdo. An expensive operating system encumbered by a proprietary commercial license was very much the opposite of what FOSS programmers and users envisioned as the best way to run a computer.

Ironically, the operating system that—more than any other software project of the 1960s and 1970s—engendered many of the practices and ideas that became central to FOSS users in later decades ended up being diametrically opposed to everything FOSS stood for. Unix's commercialization became the catalyst that pushed programmers such as Stallman and Torvalds to develop uncommercial, free clones of the operating system.

## HACKER ORIGINS: THE RECEIVED WISDOM

During the high point of the popularity of a free Unix, prior to the 1984 commercialization, programmers and computer users shared particular cultural values. This chapter has already noted that several of the salient features of FOSS culture and development strategy—such as decentralized, collaborative coding and resistance against monolithic authority—were evident in the Unix community in the 1970s and early 1980s. Those characteristics did not arise out of the ether and were not unique to the Unix world.

Instead, they reflected the values that writers about FOSS have called the *hacker ethic*. The term refers to the social, political, and cultural values that predominate within a certain segment of the computing world, including but not limited to the early Unix community. Although the hacker ethic has its roots in a period that predates the emergence of FOSS software defined as such, it played a central role in shaping how FOSS projects evolved. It also influences how FOSS developers and users think about software and its political, social, and economic significance today.

Before taking a close look at the origins of hacker culture, another note regarding language is in order. In this book, the word *hacker* refers to the class of programmers who espouse the hacker ethic—which, broadly defined, involves a commitment to creativity, exploration, collaboration, and transparency in the use of computers. The association of hackers with programmers whose chief goal is to break into computer systems toward nefarious ends is a usage that most FOSS developers and other technically inclined individuals reject and consider to be an abuse by the popular press of the term's original meaning. In other words, hackers are curious, open-minded programmers who view software development as a creative, healthy endeavor that can make the world a better place; they are not cybercriminals.[27]

Even with a clear sense of what *hacker* means for this book's purposes, it remains impossible to define the hacker ethic in a singular or precise way. Although it is easy to identify in a general sense what hackers tend to believe, hackers do not espouse a single, universal set of values. Nor did the various principles they endorse originate in a uniform setting.

Nonetheless, the two leading studies of hacker culture that focus on the history of Unix and FOSS have attempted to define a particular hacker ethic and explain its origins, although they have not arrived at the same conclusion. The first such work, *Hackers: Heroes of the Computer Revolution*, published by journalist Steven Levy in 1984, drew heavily on the accounts that Levy collected from programmers who had worked at major university computer research labs, including Stallman.[28] In Levy's telling, hacker culture was born in the late 1950s, when members of the Signals and Power Committee of MIT's Tech Model Railroad Club (TMRC)—which initially had nothing to do with the university's computer labs—assumed an interest in programming. They began taking classes in the subject that the university had recently added to its curriculum. Working on IBM 704 and 709 computers and later on a TX-0 machine, MIT students from the TMRC with a passion for tinkering and technical exploration learned to program in the LISP language. They competed to make the smallest, most efficient possible versions of software programs.[29]

At first, the creative endeavors of these young programmers were choked by bureaucratic policies that prohibited anyone other than authorized technicians from directly accessing the campus computers. Yet the students quickly learned to circumvent such restrictions by breaking rules and electronic security barriers—even constructing their own keys, if necessary, to get past locked doors—in order to use the machines.[30] These types of unauthorized entries constituted "hacking" in the jargon popular among MIT students at the time. Hence the origin of the word *hacker*.[31]

Within the eccentric community of student programmers that grew out of the TMRC's ranks, "a new way of life with a philosophy, an ethic, and a dream" coalesced, as Levy put it.[32] He summarized the core values of the nascent hacker community as follows:

> "Access to computers—and anything that might teach you something about the way the world works—should be unlimited and total."
>
> "All information should be free."
>
> "Mistrust Authority—Promote Decentralization."
>
> "Hackers should be judged by their hacking, not bogus criteria such as degrees, age, race, or position."
>
> "You can create art and beauty on a computer."
>
> "Computers can change your life for the better."[33]

For Levy, these points explained what motivated and governed the activities of the TMRC programmers and of developers such as Stallman, whom Levy called "the last true hacker."[34] Because Levy was writing just before the launch of the GNU project and the Free Software Foundation made the creation of freely accessible and shared code a conscious effort, his book was not an attempt to understand the culture of the FOSS community, which did not yet exist. Still, because *Hackers* traced a particular set of core values related to computers and programming from what Levy saw as their earliest incarnation at MIT in the late 1950s through their manifestation in the activities of Stallman and other GNU collaborators in the early 1980s, the book proved to be the first and most enduring interpretation of the values that define FOSS culture.

Levy's book formed the basis for the other major exploration of the origins of FOSS culture produced to date, Raymond's essay on "A Brief History of Hackerdom." Raymond differed from Levy on certain points. For instance, Raymond traced "the hacker culture as we know it" to 1961 rather than to the late 1950s, skipping over developments that occurred prior to MIT's acquisition of a TX-0 computer. He also afforded more weight than Levy to the role of programmers involved in ARPANET (Advanced Research Projects Agency Network), the main predecessor of the Internet, in spreading hacker culture beyond the MIT campus. Similarly, he made much of members of a group that he called the "Real Programmers," technicians who worked on the earliest computers and generated some of the oldest examples of what he called "revered hacker folklore"— even though the Real Programmers did not, in Raymond's estimation, directly precede hackers properly defined.[35]

The most important variation between Raymond's and Levy's takes on hacker culture, however, appeared in their definitions of what hackers actually believe. Levy summarized the hacker ethic in terms of the distinct points cited above, and Raymond's brief definition of the hacker ethic included only the beliefs "that information-sharing is a powerful positive good," "that it is an ethical duty of hackers to share their expertise by writing open-source code and facilitating access to information and to computing resources wherever possible" (here Raymond uses a term, *open-source*, with which some self-described FOSS hackers, particularly Stallman, take issue), and "that system-cracking for fun and exploration is ethically OK as long as the cracker commits no theft, vandalism, or breach of confidentiality."[36]

In a longer-form definition of what he called the "hacker attitude," Raymond added these points:

1.  The world is full of fascinating problems waiting to be solved.
2.  No problem should ever have to be solved twice.
3.  Boredom and drudgery are evil.
4.  Freedom is good.
5.  Attitude is no substitute for competence.[37]

Both of Raymond's descriptions of hacker values and ideals are generally consonant with the culture that surrounds FOSS today. But so is Levy's definition of the hacker ethic, even though—apart from the similarity between Raymond's declaration that "freedom is good" and Levy's that "all information should be free"—there is no direct overlap between these two writers' interpretations of what the hacker ethic involves.

Despite this major difference from Levy's account, as well as the minor variations outlined above, Raymond's explanation of the origins of hacker culture otherwise generally reiterated the story Levy had presented. And because Raymond was a prominent member of the FOSS community who was consciously writing about its history, his essay played a leading role in shaping how most FOSS programmers and users think about where their cultural values came from. That is why, for example, nonelite members of the FOSS community have described the GNU General Public License, which governs Linux and many other major FOSS projects, as a tool designed to "forc[e] users of the code to obey the hacker ethic," as one Slashdot user put it.[38] Similarly, it was not a coincidence that Pekka Himanen's book *The Hacker Ethic: A Radical Approach to the Philosophy of Business*, one of the first works to attempt to explain how the hacker ethic could apply to the business world, opened with an introduction by Torvalds, the most prominent FOSS developer of the time.[39]

As a result, the received wisdom suggests that hacker culture originated among programming students at an elite academic institution, MIT, who sought to eliminate any barriers that prevented them from using computers as creative tools. From that base, according to Levy and Raymond, hacker culture gradually spread further afield as members of the first hacker generation took jobs at other institutions and began communicating over the nascent Internet.

## DEEP HACKER HISTORY

That is a good story, and there is no reason to doubt the various facts that Levy and Raymond present. Yet there are two main weaknesses with their interpretations of hacker culture. The first is that these two writers arrived at remarkably different conclusions of what hacker values entailed. This divergence suggests that there is no such thing as a singular, specific hacker ethic to which FOSS values and practices can be directly and singularly traced.

This is to be expected, at least to an extent. Hackers are a large and diverse group. They have no official authority who can speak on their behalf or organize a formal attempt to define what they stand for. It is impossible to write a summary of the hacker ethic that accurately describes what all hackers, in all times and all places, believe. Still, it is striking that Levy and

Raymond arrived at such markedly different conclusions, which exhibit almost no common attributes, in their efforts to define the hacker ethic or hacker attitudes.

A tempting explanation for this disparity is that *hacker* is too nebulous a term. Because it can apply to virtually any programmer, it is impossible to determine who is and who is not a hacker and therefore to decide which cultural attributes to associate with hackers. That explanation is not very useful for explaining why Levy and Raymond arrived at such different interpretations of the hacker ethic, however. They both wrote about the same general groups of people and individuals, most of whom were programmers in university labs in the 1960s and 1970s. It was not because Levy and Raymond studied separate communities that they defined the hacker ethic so differently.

The second weakness with their interpretations of hacker culture is that they attempted to trace the origins of hacker culture to too narrow a milieu—one centered on MIT. They also regarded hacker values as being more novel than they actually were. As a result, they focused on overly specific examples of what hackers believe and failed to recognize the overarching values—openness and transparency—that have shaped not only hacker behavior but also much larger institutions and revolutionary movements since the eighteenth century.

The goals of increasing public accountability, resisting arbitrary authority and access controls, and rewarding individuals based on the quality of the contributions they make to the common good rather than on arbitrary characteristics such as race or social status have fueled revolutions and engendered the political norms of numerous modern societies. Against this backdrop and on these grounds, MIT undergraduates in the

TMRC justified picking door locks and breaking passwords to gain access to computers, even if they lacked permission from the authorities of the university to do so. On the basis of these principles, Unix programmers, as the earlier part of this chapter notes, collaborated across continents and shared code with one another freely, despite economic incentives against working in such a way. And these goals drive most members of the FOSS community today. They believe that software works best when the people who create and run it are free to choose which programs to use, to share code and information in ways that ensure the maximum creative potential of all parties, and to collaborate in ways that they deem most suitable rather than through the structures imposed on them by proprietary specifications and closed code.

Neither the MIT hackers of the 1950s and 1960s nor the FOSS revolutionaries who coded in their image were the first programmers to endorse these values. On the contrary, as Ensmenger has persuasively argued, since the early decades of the computer age, programmers of all stripes have tended to resist centralized control in the same way that Raymond and Levy's hackers did.[40] Similarly, Russell has shown that since the 1970s, "openness" in its various definitions has played a vital role in shaping numerous dimensions of the software and computer world.[41]

Not all programmers interpreted the importance of openness in a way that led them to share code freely. That practice was unique to the FOSS community. Yet belief should not be conflated with action. From the perspective of what programmers generally believed, not how they acted on those beliefs, the hackers whom Raymond and Levy discussed were not different from other programmers of their time. They all championed openness. FOSS developers merely sought to implement openness in a particular way, which set them apart.

## HACKER CULTURE AND ACADEMIA

Why did FOSS programmers decide that sharing code was the best way to promote openness and transparency but other developers acted differently? More than the nebulous "hacker ethic" that Raymond and Levy attempted to describe, the influence of academia on FOSS programmers provides the key to this choice.

The academic community is a diverse and nuanced place, and not all academics have endorsed values that align with those of FOSS hackers. In general, however, the academic community has long championed the sharing of information, decentralization of authority, and objective peer review, principles that are consistent with hacker values. Most universities were centers of open exchange even before the age of democratic revolutions in the eighteenth and nineteenth centuries made such values politically and socially salient. For these reasons, it was natural that programmers and computer enthusiasts who plied their trade in university labs approached code and development in a manner consistent with the principles of academic culture, even after the rise of commercial software made those practices less common among programmers as a whole than they had been in the first decades of computing.

The suggestion that hacker culture and the hacker ethic are extensions of academic practices—and that this is why FOSS culture, which was shaped by hackers, varies in many ways from the proprietary software world of commercial business interests—is not new.[42] Nikolai Bezroukov produced what remains the most explicit articulation of this idea in a 1999 article. Writing shortly after the open source software movement defined as such was born amid great fanfare on the part of people like Raymond, Bezroukov wrote that FOSS development could be best understood as "a special type of academic research" rather than something never seen before. The FOSS community, he contended, "is more like a regular scientific community than some [open source software] apologists would like to acknowledge."[43]

Raymond rejected these claims. He recognized the similarity in the values espoused by academics and hackers and the fact that many hackers had close ties to academic institutions, but he concluded that the two cultures merely "share adaptive patterns" without being "genetically related."[44] In the absence of other widely read histories of hacker culture, Raymond's criticism of the notion of a direct link between academia and the hacker community has ensured that few participants in the FOSS world today recognize how directly the cultural values that drive FOSS development descend from academic culture.

Nonetheless, the historical connections between academic institutions and hacker culture and the influence of the former on the latter are impossible to deny. First, the groups of hackers that Levy wrote about—from the initial generation of student hackers who encountered computers for the first time at MIT in the late 1950s and early 1960s to Stallman, who was on the computing lab staff of MIT until he resigned to launch the GNU project in January 1984—incubated their philosophies about technology within a setting that championed the principles of academic freedom and sharing.[45] This characteristic distinguished these hackers from many of the leading figures of the proprietary software world, which has little in common either culturally or philosophically with academia. It is likely not a coincidence that people like Bill Gates and Steve Jobs, both of whom spent only brief periods at universities prior to dropping out, created software companies and cultures that were very different from the endeavors launched by hackers and that did not value transparency, sharing, or community consensus.

Second, the Unix world—where hacker culture established some of its strongest early roots and where many of the philosophies and practices that later became central to FOSS culture had their initial incarnations—was intricately linked to academia. Unix itself was born at Bell Labs. It was not a university, but in the 1960s, the Bell Labs management endorsed a culture that was similar to that of academic institutions. According to Dennis Ritchie, Ken Thompson was able to write the Space Travel game that sparked the development of Unix in 1969 because the philosophy at Bell Labs was

> to hire people who generate their own good ideas and carry them out. ... Ken was doing something interesting that would turn out to be valuable. ... When a good university hires young professors, what do they expect them to do? Well, a certain amount of grot and service of various kinds, but really to have good ideas that somehow make an impact.[46]

Third, while neither of the founders of Unix were employed permanently in academia, Thompson retained an association with the scholarly community that was strong enough for him to take a one-year visiting professorship at Berkeley, his alma mater, in academic year 1975–76, during a formative period in the development of Unix.[47]

Finally, until Unix's commercialization in 1984, scholarly communities played an important role in constricting the ability of AT&T to centralize development of the operating system or prevent Unix users from sharing code with one another. That was because, as developer Eric Allman noted, managers at companies like AT&T worried that, if they failed to cooperate with academic developers who were working at universities, their own products would be superseded by superior alternatives written by academic programmers. According to Allman, "What happens is that industry decides 'Oh, we wouldn't want the university to have that because we might lose it'; then the university does it themselves, so industry goes 'Oh my God. We wanted ours to be standard, we'd better give it to them.'"[48] In this way, the academic community's focus on sharing and openness trumped other groups' interest in restricting access to the code that fueled the Unix hacker culture.

For all of these reasons, hacker culture is inseparable from the principles and practices of the academic community out of which it was born. It also shares strong affinities with related political, social, and economic ideologies that privilege openness, transparency, cooperation, and sharing. Its origins are thus not as narrow and specific as Levy and Raymond supposed. Nor are hackers outlying freaks who have little in common with the mainstream. That is an image of hacker culture that stereotypes have helped to construct—and which unshaven, nonconformist FOSS leaders like Stallman have reinforced—but it is a misleading approach to understanding where hackers came from or why they believe and act as they do.

## HACKER REVOLUTION?

It is possible to trace hacker culture's origins to a particular set of philosophies that were widely influential in academia and elsewhere well before the invention of modern computers, but it is difficult to think of hackers as revolutionaries in the modern sense of the word. Today, the word *revolution* implies sudden, radical change. Hackers who call themselves revolutionaries therefore suggest that the software they write was the catalyst for a major rupture. If, as I argue above, hacker values were not unique or novel to self-described hackers—but rather reflected beliefs that had been shared among programmers of all types and across academia since well before the start of the FOSS revolution—then the FOSS revolutionaries who actively promoted those values do not seem revolutionary by modern standards. They were continuing a tradition that already existed, not creating something radically new.

Yet *revolution* also has an older meaning. Before the eighteenth century, the word was associated with astronomy. It described the circular movement of celestial bodies—that is, the way they *revolved*—around a central point.[49] A revolution in this sense did not change something irrevocably but restored something to its original setting.

It might be best to think of FOSS hackers of the 1980s, 1990s, and 2000s as "revolutionaries" in terms of the earlier definition of the word, even if the hackers themselves have not necessarily done so. They were not insurgents who introduced radical innovations into the software world. They were reformers who helped return things to the way they had been. They restored the practices that predominated during the lost golden age when Unix was free, source code was regularly shared and hackers could hack unencumbered by restrictive licenses or binary-only distribution policies for code.

The beginnings of the revolution that brought things full circle—that restored them to the way they had been—are the subject of the next chapter, which examines the birth and early history of the GNU project and the Free Software Foundation. Those initiatives gave FOSS a conscious identity and meaning. They made it possible for the programmers and users who stayed true to the hacker ethic to begin envisioning a software revolution worth fighting for.

# 2 INVENTING THE FOSS REVOLUTION
## Hacker Crisis, GNU, and the Free Software Foundation

ONE OF THE DIFFICULT TASKS for historians is identifying when revolutions begin. Their starting dates tend to be defined retrospectively by the revolutionaries themselves. The townspeople and mutinous soldiers who stormed the Bastille on July 14, 1789, probably did not believe they were starting a revolution. But within several months, the Bastille attack and other upheavals around France were amalgamated within the popular imagination into a singular event that contemporaries began calling the French Revolution.[1] And through a politically inflected process of collective-memory construction, the revolutionaries singled out July 14, 1789, rather than any of the other momentous dates from that year, as the nominal start of the French Revolution.

Neither was the equivalent date in American revolutionary history, July 4, 1776, the sudden beginning of a revolution. American rebels had been killing British troops for more than a year by the time elites in Philadelphia declared independence from the British crown. Here again, the date that marks the nominal beginning of a revolution had less significance for people living in the moment than it does for posterity.

The beginnings of the free and open source software (FOSS) revolution were equally murky. No single event signaled the start of the movement. Richard Stallman's announcement of the GNU operating system in 1983 was a key development, but that initiative took considerable time to gain the momentum necessary to sustain itself. In the meantime, other changes in the Unix world, including the morphing of the Berkeley Software Distribution (BSD) into a free alternative to AT&T's Unix operating system, were equally important in promoting software freedom.

This chapter traces the numerous developments in the 1980s that helped to launch FOSS as a conscious movement. No one development on its own amounted to a revolution, but collectively, they snowballed into a broad initiative that eventually revolutionized the way computer code was written, shared, and consumed.

The chapter also evaluates what the first generation of free software revolutionaries did well and why they failed to achieve all of their goals. It highlights their engagement with political and social movements that shared certain ideological affinities with the FOSS revolution. It shows how a commitment to pragmatism ensured much success. Yet it also demonstrates the short-sightedness from which the first FOSS revolutionaries suffered in some respects, especially regarding modes of organizing software development and the significance of the emerging personal computer (PC) market.

## HACKER CULTURE IN CRISIS

By the early 1980s, hacker culture was in crisis. The hacker community, which in the previous decade had been cohesive and defiant of challenges to its mores, faced a series of obstacles that hackers felt threatened to destroy their ability to use computers in the way they saw fit.

Chief among those challenges was the commercialization of Unix. As chapter 1 explains, in the early 1980s, the companies that owned Bell Labs were reorganized. This freed AT&T from the legal interdiction on the sale of products not related to telephone or telegraph service or equipment, which it had been subject to since 1956. The company began working to turn Unix into a commercial product, which entailed raising the licensing fees for the operating system.

The practical consequences of this change should not be overstated or conflated with the close sourcing of Unix. The commercialization of the operating system did make it considerably more expensive for users to obtain a license that granted them access to the operating system's source code. In addition, beginning with the release of V7 Unix in 1979, AT&T no longer allowed universities to share the source code with students for educational purposes.[2] Yet neither of these changes turned Unix into an entirely closed source operating system. Licenses that permitted full access to the Unix source code remained available, albeit for steep prices. In the late 1970s and early 1980s, other vendors adopted policies that entailed releasing software only in binary form, as this chapter explains. The source code of their software ceased to be available to users. But AT&T Unix did not become one of these products.

All the same, the commercialization of Unix made it harder to obtain the source code, especially for hackers who did not have institutional access through an employer or university that could pay the commercial licensing fee—which exceeded

$100,000 by the late 1980s.[3] And because a source license from AT&T was legally required to run other operating systems, such as BSD, that incorporated some Unix code but included other programs that AT&T did not own, the steep licensing fees burdened users throughout the Unix community, even if they did not use AT&T's version of the operating system.

All of this was troubling for hackers. But the need to obtain a commercial license for Unix or the high cost of the license were not what truly irked them. The issue that raised hacker hackles more than any other was that Unix had become a commercial product. That change threatened to limit Unix's usability as a platform for exploration, sharing, and creativity.

It was clear that cost and source code availability were not the primary factors in hackers' anger over AT&T's commercialization of Unix because the operating system was not cheap prior to 1983. Although the six-figure licensing fees that AT&T imposed by the late 1980s raised the bar for access to Unix source code, the price for a Unix license in the 1970s, when the operating system was still a freely shared, noncommercial product, had been as high as about $20,000 dollars, hardly a negligible sum.[4]

Rather than price or access to source code, the principal threat to hackers that arose in the early 1980s was the simple idea that commercial Unix would stifle the hacker ethic. Commercialization promised to make it harder for developers to share code with one another and to collaborate in developing Unix. That change helped to engender a crisis within the community of Unix hackers.

The attitude they adopted was no surprise. Hacker antipathy toward commercial code was clear well before AT&T turned the operating system into a business product. Until the early 1980s, most Unix programmers resisted the idea of deliberately promoting the software in any way that smacked of commercial marketing, even if the goal was merely to attract more users to the community. For instance, at a January 1979 USENIX conference, hackers booed a speaker off the stage "because he was a marketing consultant or something," according to Brian Redman, one of AT&T's Unix developers.[5] Attitudes changed somewhat by 1980, when /usr/group—an organization "dedicated to the promotion of the UNIX operating system," according to its charter—was founded, introducing a conscious marketing effort to the hacker community.[6]

Yet leading Unix hackers continued after that time to express anxiety about how the trappings of commercialization might undercut the things that really mattered in operating system development—notably, creativity and exploration. Thus Unix founder Ritchie, reflecting on the operating system's history, declared in the early 1990s that a "danger" threatening "good computer science research" was the risk "that commercial pressure of one sort or another will divert the attention of the best thinkers from real innovation to exploitation of the current fad, from prospecting to mining a known lode."[7] More than concerns about money, licensing bureaucracy, or source code, the predominance of thinking like Ritchie's within the Unix community was one of the key factors that fomented hacker angst following the operating system's transformation into a commercial product.

This does not mean that Unix hackers were opposed to commercial endeavors of all types or to the introduction of cash transactions into the software realm. On the contrary, as the latter part of this chapter shows, one of the most significant innovations of the GNU project and the Free Software Foundation in the 1980s was to pioneer a way to sell support services related to free software, showing that FOSS could thrive within a commercial ecosystem even if the code itself was not for sale. GNU distributed versions of its software on tape and disk for fees that were more than nominal, although it did so as a way to support development rather than as part of a commercial venture. These practices made clear that it was not the idea of making money in the computer and software business that bothered hackers.

Instead, hackers found objectionable the practice of turning a profit by selling software toward purely commercial ends. Under these conditions, software ceased to serve the common good. It became just another way to make money, with no greater purpose. That is why commercializing code starkly contravened the principles of the hacker ethic and led Unix hackers to revolt when AT&T made the operating system part of its revenue stream, rather than a research project that pushed the boundaries of computer science knowledge.

The second major challenge to hacker culture that arose in the early 1980s was the practice of distributing software only in binary form, without source code. Binary software could run on any computer for which it was designed, but because it lacked the source code from which binaries were compiled, people using the programs had no practical way to modify them or study how they worked. The software became closed source. Interest in binary-only distribution was evident by early 1980, when analysts deemed it to be the best means of "protecting a software vendor's proprietary rights."[8] IBM brought the practice mainstream when it became the first large company to cease distributing source code with most of its products in February 1983.[9]

For most software users at the time, binary-only distribution was of little concern. By the early 1980s, the importance of software in business settings was outpacing its significance in research environments. Business owners cared little about viewing the source code for the programs they were using so long as the programs did what businesses needed them to do. If users wanted to add new features to the programs, they could request them from the vendors who sold the software, who had an incentive to implement more features in order to succeed in the increasingly competitive commercial software market.

Hackers, however, thought differently about all of this. Closed source software was at odds with the core values of a generation of hackers for whom access to source code—and, by extension, the transparency of that code and the ability to modify it and share ideas freely—was an intrinsic part of what it meant to use a computer. For that reason, the growing adoption of binary-only distribution practices by many of the organizations developing software, even in places such as universities, by the early 1980s presented another major crisis for the hacker community.

Software had been born free, it seemed to hackers. Yet everywhere it was shackled in chains.

### MICROCOMPUTERS AND THE INTERNET

The commercialization of Unix and the advent of closed source software were the greatest challenges to the hacker ethic at the time. But other changes negatively affected hackers as well. Although other developments did not undercut hackers' goals as directly as the two factors discussed above, they nonetheless helped to push hackers from the center of the computing world (where they had been in the heyday of noncommercial Unix) to its margins.

The rise of microcomputers was one such change. The Unix hacker community had been born and thrived on large, institutional computers, especially the PDP line from Digital Equipment Corporation and later VAX hardware from the same company. Those types of machines remained readily available into the late 1990s. In terms of market share and public mind share, however, they were eclipsed starting in the late 1970s by smaller personal computers. Apple began selling microcomputers in 1976. Commodore released the PET PC computer in 1977, the same year the Tandy TRS-80 hit the market. Matters came to a head in 1981, when IBM introduced its first PC for the mass market, the 5150. Based on an Intel microprocessor, the IBM PC helped set a standard that other computer manufacturers rapidly emulated, building clones of the IBM PC that could run the same software. The result was a thriving new market for PC hardware and software.

The PC revolution was not a necessarily bad thing for hackers. In a way, it was a welcome development because PCs made computer hardware easier and more affordable than ever before for individual consumers. However, it did nothing to help the cause of keeping Unix free of commercial entrapments or otherwise protecting hacker culture from the crisis it faced by the early 1980s. PCs were designed and sold almost exclusively for commercial purposes, lacked sophisticated hardware, and were generally able to support only a single user. They were a far cry from the powerful, research-oriented computers on which Unix was born. Although a handful of third-party ports of Unix for PC hardware emerged starting in the early 1980s, including most notably Coherent from the Mark Williams Company and Microsoft Xenix, they were as commercialized as AT&T's Unix.[10] They did little to advance the cause of hackers who wanted to keep Unix free as momentum in the computing industry shifted toward PCs.

Another change that affected hackers in a serious way by the 1980s was the growth of the Internet. The World Wide Web was not conceived until the end of the decade, but throughout the 1980s, computer network access remained limited for most people, and many of the protocols that became foundational to the Internet in the 1990s were still in development. Yet the increasing use of tools such as email and Usenet newsgroups and the distribution of software over the Internet changed the stakes of computing and introduced new paradigms into the hacker culture.

These developments benefited hackers in many ways, not least by making collaboration and the sharing of information easier and faster. But they also opened new opportunities for people who called themselves hackers to break into computers remotely or steal data over the network. Most hackers did not condone such activities. But the fact that a few individuals engaged in them in the name of the hacker community sullied its image, giving rise to the negative sense of the word *hacker* that predominates in general usage today. As a result, being a "true," vocal hacker became more difficult as the Internet age dawned during the 1980s.

The marginalization that the personal computing revolution and the rise of the Internet imposed on hackers perhaps explains why they did not take greater advantage of these new resources at the time. In many ways, as this chapter details below, the hacker community of the 1980s, which coalesced around Stallman's GNU project, avoided focusing its energies on microcomputers. Not until after it became impossible to ignore the importance of PCs did GNU finally begin releasing software for them in the early 1990s. Meanwhile, the GNU hackers also made less use of the Internet as a tool for collaboration between developers than they might have, preferring instead to centralize development in the environs of Cambridge, Massachusetts.

As the latter part of this chapter contends, both of these policies stunted the momentum of the GNU project. They also help to explain why some of GNU's efforts were superseded in the early 1990s by programmers in the Linux community who, as chapter 3 shows, took microcomputers seriously and made widespread use of the Internet.

We will tackle these issues soon enough. For now, let us explore how hackers confronted the threats they faced in the 1980s.

## BERKELEY VS. AT&T

Hackers responded to threats of commercialization and closed source software in the 1980s in three main ways. First, programmers at the University of California, Berkeley, eventually created a Unix clone that was free of commercially licensed code and suited the needs of many hackers. That initiative started in 1986, when a programming team led by Ken Bostic began working to disentangle code derived from AT&T Unix from BSD, the enhanced version of Unix that Berkeley developers had been writing and distributing since 1978. Bostic kept the Unix community up to speed regarding the progress of the effort by announcing at USENIX conferences which percentage of BSD programs were free of AT&T-owned code.[11]

By June 1989, the Berkeley team had produced enough freely licensed software to issue its first independent software release.[12] Called NET 1, the platform consisted mostly of code related to networking. It was a far cry from a full Unix implementation. Still, NET 1 was important because organizations could legally use it without obtaining a commercial license from AT&T.[13] That was a big deal at the time.

BSD became a bigger deal in 1991, when the Berkeley developers took the momentous step of issuing a complete, standalone Unix-like operating system. They called it NET 2 (even though it contained much more than networking code).[14] Accompanied by a derivative port for microcomputers based on Intel 386 and 486 processors called BSD/386, which was available with full source code for about $1,000, NET 2 was a noncommercial Unix clone that implemented all of the important functionality of Unix itself and yet freed hackers from having to work with AT&T if they wanted to run Unix.

Unfortunately for Unix hackers, the Unix clones from Berkeley appeared too late to provide an easy transition away from AT&T's Unix. Had NET 2 arrived earlier, it might have obviated the need for Richard Stallman to launch the GNU project described below—especially if the BSD licenses had proved more satisfying to Stallman. But by 1991, it was too late for BSD to become the central rallying point for all hackers who were nostalgic for the days of noncommercial Unix.

In addition, beginning in 1992, legal challenges against Berkeley's software from Unix Systems Labs, the company that took over ownership of Unix from AT&T in 1990, stunted the adoption of BSD/386 in serious ways. Although these lawsuits, which the next chapter describes in detail, eventually were resolved in Berkeley's favor, the uncertainty over BSD/386's liability restricted its widespread adoption and made it a poor substitute for free Unix in hackers' eyes.[15] So did the fact that although Berkeley's software counted as free and open source by most modern definitions, it was governed by permissive licensing that allowed users to modify the code and distribute binary-only versions of it. As a result, the BSD licenses provided no assurance that the code from Berkeley would promote continued collaboration and transparency within the Unix community.[16]

## THE OPEN SOFTWARE FOUNDATION AND SOFT-WARE STANDARDS

Another way that hackers confronted the threats they faced from commercialization and closed source software in the 1980s emanated from the Open Software Foundation. Formed in May 1988 by a group of companies with support from some hackers, this organization was a response to AT&T's purchase of a large share of Sun Microsystems, a major manufacturer of computer hardware, in late 1987. Sun subsequently announced that it would jettison the BSD-based version of Unix that it previously shipped with its computers in favor of AT&T's Unix. The move spawned worries within the industry that AT&T and Sun would collude to make their version of Unix incompatible with BSD and other variants in order to gain an edge in the market.

The Open Software Foundation sought to forestall that eventuality by promoting a set of software standards across Unix versions. The standards would ensure that Unix programs remained compatible between different platforms and variants of the operating system. In addition, the organization planned to build its own hardware-agnostic variant of Unix to compete with AT&T Unix in the market.[17]

Sharing source code and combating AT&T's high licensing fees were not major concerns for the Open Software Foundation, which formed at a time when the word *open* referred to software standards, not source code. As a result, although the idea for the organization was first proposed by Armando Stettner, one of the old-guard Unix hackers, other segments of the hacker community expressed little enthusiasm for the initiative. In their June 1988 newsletter, the GNU developers declared that the Open Software Foundation "saddened" them because the organization's goal was "to develop yet another proprietary operating system" rather than one that GNU hackers viewed as a satisfying replacement for AT&T Unix.[18] The Open Software Foundation's donation of $25,000 to GNU shortly afterwards did little to sweeten the initiative's image in hackers' eyes.[19]

Ultimately, the Open Software Foundation proved to be of little real importance for the Unix world. The initiative began fizzling by the end of 1993. By that time, AT&T had sold its interest in Sun, and a poor economic climate weakened the resolve of the companies that had supported the Open Software Foundation. The organization disbanded entirely by 1996.[20]

Yet although the Open Software Foundation proved relatively short-lived and failed to secure the endorsement of a large segment of the hacker community, it was important nonetheless for the history of FOSS. It was one of the first organizations to make the promotion of open standards—which played a major role in FOSS's success during the 1990s, especially in the Internet market—an explicit part of its agenda. The organization also attracted greater attention to the ways in which AT&T's interest in aggressively selling Unix as a commercial product could stifle the free sharing of software. Hackers shared that concern, even if, on the whole, their opposition to AT&T's Unix policies was very different from that of the companies that organized and funded the Open Software Foundation.

## GNU: ONE MAN'S CRUSADE TO "FREE UNIX!"

The third major response of hackers to the threats from commercialization and closed source software that they faced in the 1980s—and the one that proved to be the most enduringly important for the history of FOSS—was the GNU project, which Stallman announced in 1983 and launched in early 1984.

Today, GNU is one of the biggest names in the free software world (although the project does not carry equal weight in the open source ecosystem because its founder and most of his close collaborators have fiercely rejected use of the term *open source* to describe their work, as chapter 5 explains). Yet in GNU's infancy, it was not at all clear that the obscure initiative, whose membership at first consisted solely of its enthusiastic but abrasive founder, was destined for such enduring and widespread success.

Because so much of GNU's early history hinged on the activities of Stallman himself, it is worth taking a look at the man as a person—and not merely the radical software ideologue that he is often made out to be—in order to understand GNU's origins. A self-described "weird" child, Stallman grew up splitting his time between two single-parent households in New York City.[21] He has identified himself as "borderline autistic," a trait that perhaps did much to incline him as a teenager to recreate in the isolated environment of computer labs.[22] Beginning in the summer following his junior year of high school, when he was hired by the IBM New York Scientific Center in Manhattan, Stallman started earning money for his work on computers.[23]

The importance of software to Stallman's life increased further when he enrolled at Harvard in 1970. By the end of his freshman year, he began visiting the Artificial Intelligence (AI) Lab at nearby MIT, where the resident hackers—who, in Stallman's words, were "more concerned about work than status"— welcomed him even though he was an outsider to the institution and a mere undergraduate. They soon offered him a programming job.[24] For Stallman, the reception he enjoyed at the AI Lab exemplified everything that was great about hacker culture and the community that surrounded it. At MIT, that culture was sustained by the old guard of first-generation hackers.

Unfortunately for Stallman, that generation of hackers was growing old by the time he arrived in Cambridge. Its ranks were already thinning as more and more programmers adopted the view that using computers as vehicles for joy and creativity undercut their potential to contribute to ostensibly more serious purposes. By the mid-1970s, the hacker community at MIT was under attack from programmers such as Joseph Weizenbaum, who complained about student hackers' "rumpled clothes, their unwashed hair and unshaved faces," which "testify that they are oblivious to their bodies and to the world in which they move."[25]

Perhaps because Stallman was wary of falling into the ranks of programmers like Weizenbaum, in the fall of 1975, a year after he received a bachelor's degree from Harvard and entered the Ph.D. program in physics at MIT, he dropped out of school, never to return. He began working at MIT as a full-time programmer at the AI Lab, one of the few places where hacker culture remained alive and well at the time.

Yet even the AI Lab could not long withstand the changes afoot in the software world. By the late 1970s, funding from the U.S. Department of Defense, which had sustained much of the early computer science research at MIT, was withering. In its absence, the AI Lab sought out partnerships with private investors that were interested in commercial software development. By 1980, a majority of the lab's programmers, including a number of those who had formerly been free-spirited hackers, were devoting a large share of their time to commercial software projects. The most egregious blow to the hacker community at the lab arrived in 1982, when administrators opted to install a commercial operating system, Twenex, on the lab's hardware, consigning its free predecessor to oblivion.[26]

For Stallman, the changes at the AI Lab compounded deeply personal frustrations with commercial software whose source code was not publicly available. Around 1980, Stallman's patience was tested by a Xerox printer at the AI Lab that had a tendency to jam, delaying print jobs until a human happened to walk by and realize that the printer required attention. Seeking to modify the printer software so that it would send a message to users at their computer terminals whenever a jam occurred, Stallman found that the software had been made available to MIT only as machine code. Without access to the source code, Stallman could not add new features to the software.

Stallman initially resolved the issue by obtaining the printer software source code from colleagues at nearby Harvard. At the time, Harvard retained a policy requiring vendors to supply the source code with any software installed on university computers. With the source code in hand, Stallman was able to enhance the printer software in a way that made it more useful for him.[27]

Yet that happy outcome did not endure. A short time later, Stallman requested a different version of the Xerox printer source code from a programmer at Carnegie Mellon University, another early bastion of hackerdom. That programmer, whose name Stallman no longer recalls, refused to share the source code, which had come under the protection of a nondisclosure agreement that Xerox had adopted to help commercialize its products. The Carnegie Mellon programmer's refusal to share the code made Stallman "so angry I couldn't think of a way to express it," he recalled. "So I just turned away and walked out without another word. … I was stunned speechless as well as disappointed and angry."[28]

The disintegration of the hacker community within the AI Lab, the growth of commercial software, and programmers' refusal to share code with one another confirmed for Stallman that the hacker world faced an existential crisis—that the "Garden of Eden," a term he has used to describe the AI Lab during its heyday, had withered and might disappear forever.[29]

At its core, the issue for Stallman was not that moneyed interests had been introduced to the software world or that unfree code made it harder to implement features (such as printer-jam alerts) that improved computer users' experience. It was rather that the new norms of software development and distribution were forcing him "to betray all my colleagues."[30] When he and other hackers faced difficult choices between sharing code with one another or respecting companies' wishes, the communal spirit that had once bound hackers together through the common values of transparency, openness, and collaboration disappeared. Their disappearance left hackers like Stallman without a community. Resurrecting this community and reinvigorating the hacker culture were the goals that inspired Stallman's revolution against proprietary software.

At first, the revolution was a lonely one, though not by design. On September 27, 1983, when Stallman announced to the world—or at least to the members of the net.unix-wizards and net.usoft newgroups on Usenet, a messaging forum popular among programmers—that he intended "to write a complete Unix-compatible operating system called GNU (for Gnu's Not Unix) and give it away free to everyone who can use it," he eagerly sought the collaboration of like-minded programmers.[31] Yet in a reflection of just how much the hacker culture had ebbed since the high-water mark of its golden years in the 1970s, no one signed on to join the project at the time.

In part, the lack of enthusiasm for GNU (officially "pronounced as one syllable with a hard g" but commonly rendered "goo-new" as well) when Stallman announced the initiative stemmed from fellow programmers' wariness over its feasibility.[32] Writing a clone of Unix—an operating system that had taken a large community of programmers more than a decade to develop—was no simple proposition. As the leader of a Unix user group from the time explained, the reaction to Stallman's announcement was that "that's a great idea" but he first needed to "show us your code. Show us it can be done."[33]

GNU also suffered in its early days from a lack of a clear, inspirational vision. Today, Stallman, his fellow GNU programmers, and the Free Software Foundation articulate such decisive and strong viewpoints that it can be difficult to appreciate how long it took their ideology to develop. As Stallman himself has noted, however, "several of the philosophical concepts of free software were not clarified until a few years" after he announced the GNU project in September 1983.[34] As a result, the original announcement said nothing about, for example, free software licenses. Those became one of the most potent weapons in GNU's arsenal starting in the later 1980s. But at GNU's outset, Stallman was focusing on software, not bigger political and legal issues.

Similarly, Stallman in 1983 did not mention source code or otherwise define what he meant when he wrote about "free software." As a result, it was easy for Usenet readers to assume that he simply wanted to create software that users could share with one another without cost. Such readings missed the much more important goal of protecting the openness of source code. At the time of GNU's birth, even Stallman himself could not clearly and fully articulate what it meant for software code to be free—perhaps because the alternative, closed source software, remained a relatively novel concept at the time within the research community of which Stallman was a part.

The GNU announcement did hint at the ideological motives behind the project by mentioning Stallman's goal of preserving his "Golden Rule," which "requires that if I like a program I must share it with other people who like it. I cannot in good conscience sign a nondisclosure agreement or a software license agreement."[35] Nonetheless, lacking a clearer and more powerful explanation of the rationale behind GNU or why it mattered for hackers, the announcement failed to inspire a large following.

Stallman's GNU announcement may have lacked inspiration, but he did have a general idea of what he wished the project to do—to "Free Unix!," as he declared in the first line of the Usenet post announcing the initiative. More specifically, he envisioned creating a Unix-like kernel, "plus all the utilities needed to write and run C programs: editor, shell, C compiler, linker, assembler, and a few other things" necessary to build a clone of a Unix system. After those basic pieces were in place, the GNU team would develop programs to make its Unix-like operating system useful in real-world settings, such as spreadsheet software, text-formatting tools, games, and "hundreds of other things."[36]

Stallman envisioned the GNU system as more than a mere replacement for Unix. He wanted it to be a better Unix. It would boast features such as "longer filenames, file version numbers, a crashproof file system, filename completion perhaps, terminal-independent display support, and eventually a Lisp-based window system through which several Lisp programs and ordinary Unix programs can share a screen." Although most of these features are now so common on Unix-like operating systems that it is hard not to take them for granted, Unix in 1983 included none of them.[37] By seeking to implement them in GNU, Stallman provided an early example of FOSS programmers' belief that, for FOSS to thrive, it needs not just to emulate the competition but to do everything the competition does and more and in a way that is technically superior and enhances usability.

In addition to identifying the software programs he wanted to include in GNU, Stallman in 1983 listed the material resources he requested to make them a reality. His top priorities were donations of money and computers. In a reflection of the GNU project's humble origins in Stallman's Cambridge apartment, he specified that any computers given to the group "better be able to operate in a residential area, and not require sophisticated cooling or power."[38] He also requested donations of software from programmers who could create standards-compliant clones of Unix utilities for use as part of the GNU system.

## NASCENT GNU

Some authors of FOSS history, notably Raymond, have written that "GNU quickly became a major focus for hacker activity."[39] In fact, the project grew slowly following Stallman's announcement of the endeavor in September 1983. Although he originally envisioned beginning work on the GNU system by Thanksgiving, he did not start executing his plan until January 1984, after quitting his job at MIT. During GNU's first year, he recalls only "a few others" joining him.[40] Among them was Dean Elsner, who arrived in September 1984 "and ended up staying most of a year and writing the GNU assembler, gas," according to Stallman.[41] Stallman himself spent part of the first year of the project working as a software consultant for a British company, helping to pay his own bills while also acquiring cash for his nascent free software crusade. Such work was distracting from GNU's goals but necessary because Stallman lacked a steady source of personal income and, until that time, had enjoyed little success in securing financial donations for GNU.

At first, Stallman attempted to make the most of GNU's meager resources by finding code that he could borrow from other programmers and incorporate into the GNU system. That approach seemed more efficient to him than embarking on the more time-consuming process of writing GNU code entirely from scratch. In reality, however, the strategy led only to frustrations. To create the GNU compiler, a tool that turned source code into machine code, Stallman initially attempted to appropriate code from a compiler called VUCK. The name was an acronym for a Dutch term that translated to "Free University Compiler Kit." Stallman assumed that the presence of the word "Free" in the title meant the compiler's code was freely available to anyone who wanted it. (Presumably, he intended to give the program a new name whose acronym would be less offensive to English speakers.) Stallman was dismayed when the developer of VUCK told him the software was not free; only the university mentioned in the program's name was free.[42]

Disappointed but not yet disillusioned, Stallman next attempted to adapt a Pastel compiler that had been written at the federal government's Lawrence Livermore National Laboratory for use with GNU. He abandoned that effort, however, when it became clear that the compiler's design made it difficult to modify it for use with a Unix-like operating system.

These setbacks early after GNU's launch disenchanted Stallman, who wondered whether a more realistic approach to combating unfree software would be simply to "find a gigantic pile of proprietary software that was a trade secret, and start handing out copies on a street corner so it wouldn't be a trade secret any more."[43] That strategy, he has half-jokingly suggested, might have proved more effective than attempting to write a free operating system from the ground up.

Yet Stallman persevered along the hard road to software freedom. Following his failure to obtain a ready-made code base for the GNU compiler, he shifted attention to the somewhat simpler task of adapting Emacs, a popular text editor he had helped develop at MIT's AI Lab, for use with GNU. Here again, however, Stallman was at first thwarted by the confines of proprietary software and by his failure to appreciate how pervasively commercial software had undermined hacker culture in a few short years.

Rather than writing an Emacs implementation for GNU from scratch, Stallman borrowed code from a program called Gosling Emacs, also known as Gosmacs or gmacs, which James Gosling had created as a graduate student at Carnegie Mellon in 1981. Because Gosling allowed the free distribution of the program when he wrote it and a developer who had worked with Gosling told Stallman that borrowing the code for use in other projects would not pose problems, Stallman believed he could incorporate Gosling's code into GNU's Emacs implementation without issue. That did not prove to be the case, however. Uni-Press, a private software company to which Gosling had sold Gosmacs by 1984, threatened legal action against GNU when it learned of Stallman's intentions.[44] Once again, a shortcut to realizing the GNU vision proved impassable.

Eventually, Stallman resolved to do things the long and hard way. He wrote his own version of Emacs from scratch.

Devoid of unfree code from UniPress or anywhere else, GNU Emacs in 1985 became the first program that GNU successfully released.[45]

As a mere text editor, GNU Emacs was only one small part of a Unix-like operating system. And it was not even the most important part. A kernel, compiler, and assembler, none of which GNU possessed in 1985, were much more essential for building a working operating system.

Yet the symbolic significance for the GNU project of the Emacs release in 1985 greatly exceeded the actual functionality that the program contributed to the realm of free software. That is because GNU Emacs meant that Stallman finally had code to show to other hackers. With Emacs, he could convince them that his GNU vision amounted to more than an ambitious young programmer's pipe dream.

**GNU GROWS**

With code in hand, GNU finally gained momentum. More programmers fell in line behind Stallman, and organizations increased donations of money and equipment.

Initially, the growth was by no means explosive. GNU's total budget in 1985 amounted to around $23,000, about 80 percent of the mean annual household income in the United States in that year.[46] That was hardly the level of expenditure one would expect of an organization building an operating system that would eventually help to power millions of computers worldwide. Similarly, in another reflection of GNU's measured rate of growth, records show that only six collaborators signed copyright papers to work with GNU in 1985.[47]

The situation improved the next year, when the number of GNU copyright contracts increased to sixty-six—although some individuals signed more than one contract, meaning that GNU had fewer than sixty-six total collaborators at the time.[48] By February 1986, a list of people who were significantly involved in the GNU project included ten names, Stallman's among them. Only one of these individuals, however, was a paid GNU employee.[49] In addition, although GNU reported having received by February 1986 "one hundred responses" to requests for financial donations—thanks especially to a column in *Byte* magazine by Jerry Pournelle that promoted Stallman's project—the organization was still in great enough need of cash that GNU organizers shamelessly reiterated "money" five times on a nine-item list of requested donations in their newsletter that year.[50]

Yet within the realm of programming, GNU's expansion during its first years was more impressive than the sluggish growth of its personnel or the lightness of its coffers might suggest. By early 1986, GNU Emacs was already in use on several variants of BSD version 4.2, as well as some editions of AT&T Unix. More notably, the Berkeley hackers agreed to ship Emacs as part of BSD 4.3, which they released in June 1986, and Digital Equipment Corporation "expressed an interest in distributing" the editor as part of the software package that it provided for computers it manufactured.[51] The endorsement of what was then GNU's flagship software product by such big names in the Unix world was a major marker of success, especially since the project's Emacs implementation was barely a year old at the time.

Emacs was only one of several major programs that GNU was producing by early 1986. In February, GNU developers also reported having completed a shell, a program called gsh that was in testing and was intended for interacting with a computer. Ultimately, as noted below, gsh development faltered, and GNU opted to use the Bourne Again Shell (bash) instead. Work on a GNU C compiler—the crucial program that Stallman had vainly tried to adapt from a free third-party version—was in progress under the direction of Len Tower, whom GNU was paying a full-time salary. A handful of clones of basic Unix utilities, including "ls" (a program that lists the contents of a directory in a Unix shell) and "make" (which assists in compiling software) were already complete, and GNU developers were making progress on others. The gas assembler that Elsner had been writing since 1985 was "mostly finished"—complete with the crucial feature of being able to prepare software code for compilation on different types of hardware platforms, which meant GNU could be a portable operating system. Stallman himself was working on a debugger, which programmers use to test and troubleshoot software code.[52]

This assortment of programs and tools, only a minority of which were actually ready for use, by no means amounted to a complete Unix-like operating system. Yet they were remarkable achievements as the fruits of a project that was little more than two years old and whose growth had been virtually anemic for the first half of its history.

That progress helps to explain how GNU began attracting more outside collaborators for development and distribution of its software. Starting in 1988, GNU developers were not just cooperating with their counterparts at Berkeley who were working to make BSD free of AT&T code but were actively coordinating with them. GNU announced in February 1988 that "the next release of Berkeley Unix may contain Make, AWK and SH from the GNU project instead of those from Unix." These plans meant that GNU was "coming to Berkeley's aid," in Stallman's estimation.[53] They highlighted how the growing array of free GNU software programs was playing an increasingly important role in enabling BSD developers to excise AT&T code from their operating system. Meanwhile, GNU by early 1989 was broadening its reach to support programmers who worked on GNU software from remote sites, even though the organization initially asked all of its collaborators to spend most of their time near its headquarters in Cambridge.

Sympathizers with deep pockets—or at least with wallets fatter than that of the underemployed hacker who had launched the project—took note of GNU's progress by the late 1980s. The only significant donation Stallman had secured by the time he began work on GNU in 1984 was a single computer. By January 1987, however, Lisp Machines, Inc., a company founded by first-generation MIT hacker Richard Greenblatt, was providing free office space for the project. Six months later, GNU was receiving gifts of computer hardware, triple-digit corporate cash donations, and an answering machine.[54]

The magnitude of cash donations to GNU reached new heights in February 1988, when the project announced a $10,000 gift from Software Research Associates, a Japanese company that also committed to donating another computer to GNU. Those donations came thanks to the influence of Kouichi Kishida, who hoped GNU might provide a free Unix clone that would promote use of Unix-like software in Japan.[55] A year later, Hewlett-Packard outdid GNU's Japanese supporters with a $100,000 donation to the project as part of a program intended to increase the company's appeal in the academic market.[56]

GNU's developers did not rest on their laurels, however. Despite the sizable cash donations the project had secured by 1988, its members continued to request financial gifts to sustain GNU's rapid growth. The biannual GNU newsletters frequently reminded readers that donations to the project were tax-deductible and would be used to hire more programmers to write GNU software and create documentation for its programs, a vital resource for ensuring their widespread usability.

## FREE SOFTWARE FOR SALE

At the same time, the project generated revenue by charging fees to distribute GNU software on tapes and, later, disks. GNU code was always freely available for download without cost from servers on the Internet, and GNU developers encouraged users to copy and share the project's software directly among themselves. But the team also offered its products through the mail via tapes, the most common medium for transferring electronic data at the time. In addition to providing a convenience for users who did not have easy access to the Internet or the ability to copy media on their own, tape distributions of the GNU software suite—which the organization offered in January 1987 for $150 plus an additional $15 for the accompanying documentation manual—provided a way for individuals to contribute money to GNU while receiving something in return. Institutional users who could count the GNU software as a professional expense reimbursable by their employer enjoyed a particularly pain-free method of supporting GNU financially.

GNU maintained its tape- and disk-distribution service for several years, eventually expanding it to offer different packages of GNU software at varying price points. By the early 1990s, the most expensive GNU product was the "Deluxe Distribution" of GNU software and manuals, which cost $5,000.[57] GNU's success in enticing users to purchase tapes and disks, even though all of the software they contained was available through other means perfectly legally and at no cost, showed that it was possible to convince people to pay money for free software—especially if purchasers gained convenience or add-ons, such as documentation manuals. In that respect, the distribution service that GNU introduced in the 1980s set an important precedent for the FOSS world, where many business models today revolve around selling free or open source code in value-added form.

In a similar vein, GNU pioneered modes of free software development and distribution around which entrepreneurs were able to build a successful commercial ecosystem. GNU's first newsletter, which appeared in February 1986, included an essay by Stallman that explained how free software programmers could make a living even if the code they wrote cost nothing. Some of his proposals, such as having programmers live off of donations or having free software user groups contract with programmers to write software for them, seem simplistic from today's perspective. Yet he also suggested that free software programmers might sell "teaching, hand-holding and maintenance services," an activity that sustains a great deal of FOSS-related business today.[58]

By June 1987, Lisp Machines, Inc. was donating $200 to Stallman personally for each copy of GNU Emacs that the company delivered to customers. Because Stallman received the donations for his role as the programmer of Emacs rather than GNU's founder, this activity "proves it is possible to make a living from writing free software," GNU told its supporters.[59]

Yet the most momentous development for the commercialization of the free software space came in November 1989, when John Gilmore, Michael Tiemann, and David Henkel-Wallace founded Cygnus Solutions, the first company to build a business model that centered on providing support services for free software.[60] Cygnus grew rapidly, increasing its bookings from $725,000 during its first full year of operation to $5,700,000 five years later. In 1997, the company secured venture funding, and in 1999, during the heady days of the commercial FOSS explosion (discussed in chapter 4), it merged with Red Hat.

Cygnus was only one of several companies to build business models around free software early on. Others cropped up by the early 1990s in places as far from Massachusetts as Russia and Japan.[61] Taken together, these ventures proved that free software was by no means at odds with commercial success.

They also undercut the notion that hackers like Stallman were driven by an anticapitalist mindset. That is a tempting interpretation, especially because the commercialization of Unix had been a major factor in the decline of hacker culture and because the GNU developers emphasized assuring users that their software would always be available free of charge. Yet such explanations fail to account for the success of companies like Cygnus and grossly misinterpret the ways in which hackers such as Stallman thought about business and commercial success. As commercial software companies began proliferating in the 1980s, making money in the software world or anywhere else was not what upset Stallman and most of his fellow hackers. What sparked the GNU revolution was the breakdown of a culture in which programmers shared source code among themselves freely. In many cases, that disintegration happened because corporate managers believed they could make more money selling software if they distributed it only in binary form. Yet it was not the fact that companies were making money that upset hackers. After all, many of them worked for companies in the software business. It was rather that source code's openness had become a casualty of the business model that most companies adopted.

## EXPORTING THE REVOLUTION

As GNU's commercial significance expanded along with its suite of software programs, the project's international stature also grew. By 1988, the organization already had admirers in Japan who, as noted above, gave GNU its first large cash donation. In 1989, an anonymous supporter from England donated another $100,000 to the project, a few months after Hewlett-Packard's gift of the same amount.[62] Two years later, months before the dissolution of the Soviet Union in December 1991, GNU reported that it had "grown a branch in Russia," where a start-up company was employing ten programmers to work on GNU software. The company's goal was to sell computers that ran the operating system that GNU was creating. GNU generated enough interest in Russia by 1993 for the Society of Unix User Groups, the Russian Center for Systems Programming, and the International Center for Scientific and Technical Information to host a conference in Moscow dedicated to GNU and related topics, including the "relevance of free software to modernization and democracy in Russia and other parts of the former Soviet Union."[63] In another corner of eastern Europe, a project called "Free Unix for Romania" began distributing GNU software to Romanians in 1992.[64]

During its first decade, the followings that GNU gained outside of the United States, even though the project was founded and based in Massachusetts, showcased free software's ability to cross political and cultural borders freely. That strength continues to distinguish FOSS from many proprietary software products today, which are often not as readily adaptable for different language groups and are sometimes subject to resentment by users who view software exports from companies based in the United States as a vehicle of cultural imperialism.

In certain areas, GNU lagged during the otherwise lively years of the late 1980s and early 1990s. For example, the gsh shell that GNU developers reported as nearly complete in February 1986 never materialized because its "author made repeated promises to deliver what he had done, and never kept them," the project reported.[65] Not until the summer of 1988 did GNU complete a free shell program under the direction of Brian Fox. Called the Bourne Again Shell (bash), the software was a replacement for the Bourne Shell, a common program on Unix systems at the time. Today, bash remains one of the most popular shells on free and open source operating systems.

GNU also perennially struggled in its efforts to produce adequate documentation for the software it was creating. For many programmers, writing code is more exciting than writing clear, concise descriptions in natural language that explain how to use the code. Yet because even other programmers rarely can understand how a particular software program is supposed to work simply by glancing at its code or executing machine code created from it, software that lacks sufficient documentation is difficult to use in a serious way.

By February 1986, GNU had developed tools to help produce documentation files for its software, but its repeated calls for volunteers to create the documentation elicited few responses. By February 1988, the lack of documentation had become sufficiently acute for GNU to list its desire "to hire somebody to write documentation!!!" as one of the main reasons it sought further cash donations. Yet it was not until later that year that the documentation drive enjoyed greater success, after Dick Karpinski, a professor at the University of California at San Francisco, announced a $1,000 cash prize for writing documentation for a particular GNU program. A candidate quickly produced a manual in response to the prize offering, helping to expand GNU's lackluster documentation library.[66]

Eventually, GNU developers succeeded in writing sufficient documentation for virtually all of the programs and utilities that the project produced, a fact made clear by the rich documentation files that ship with most GNU/Linux operating systems today. GNU's struggles to generate adequate documentation during its early years, however, underscored that hackers found it much easier to write free code than to complete the auxiliary tasks that are necessary to ensure the success of free software.

## BEYOND CODE:
## THE FREE SOFTWARE FOUNDATION AND THE GPL

Producing documentation was only one of the challenges that GNU faced beyond the realm of coding. Gradually, Stallman and his collaborators recognized that if GNU was to affect how people used technology, they needed to produce much more than code or documentation. Their endeavor also demanded political and legal tools that would promote and protect the software they produced.

Stallman took the first step toward establishing a broader footing for GNU on October 4, 1985, when, ten months after he had launched GNU, he founded the Free Software Foundation.[67] As Stallman remains keen to point out today, "GNU is not a movement. It's an operating system."[68] In contrast, the Free Software Foundation provided institutional grounding that allowed Stallman, the organization's initial president, along with the fellow hackers who rounded out its board to make GNU part of a larger initiative, with a purview that extended beyond creating a Unix-like operating system.

This did not mean, however, that the GNU project avoided directly engaging with issues that did not relate to code. In June 1987, GNU developers railed against a proposed law that would have required manufacturers of electronic audio hardware in the United States to install equipment that prevented the copying of cassette tapes or the recording of music from the radio onto them.[69] That campaign represented the first time that GNU addressed an issue that involved the freedom of another technological and cultural medium, with no direct bearing on the GNU operating system.

In subsequent years, GNU promoted such endeavors as the Open Book Initiative, which aimed to distribute books and other written materials over the Internet, and the Universal Index, a project to build a database of copyright-free information.[70] In the early 1990s, GNU endorsed Project Gutenberg, a major hub for distributing free electronic texts. By 1995, GNU developers were denouncing attempts by the United States government "to ban messages that it cannot read" as a result of electronic encryption.[71] Actions like these made clear that the GNU and Free Software Foundation teams came to understand their work as having to do with promoting open culture and free society as well as free code. In this sense, they were pioneers of the debates over open access that continue into the present, as Gary Hall's work shows.[72]

The most crucial achievement of GNU and the Free Software Foundation from the perspective of free software itself, however, involved licenses and copyright law. The GNU General Public License (GPL) constituted a major innovation in the way that programmers and the general public thought about the roles played by licenses and copyright restrictions in the world of software. It illuminates more clearly than any of GNU's other initiatives just how radically the project affected the world as a whole.

At first, hackers like Stallman viewed copyright as an evil with no positive potential. As noted above, Stallman had declared when he announced GNU in 1983 that "I cannot in good conscience sign … a software license agreement."[73] Four years later, the June 1987 GNU newsletter called copyright "a public nuisance that the public tries to ignore."[74]

Such attitudes were no surprise. Beginning in the 1970s, the introduction of copyright into the world of software had been part of the campaign by commercial software companies to change the way software was written and distributed. Software copyrights came as a major shock to hackers. During the first decades of computing, the prevailing attitude toward code in legal contexts was to treat it as an intangible object rather than as something one could hold in one's hands. True, you could print it out if you wanted, but few people did that. Code existed mostly in ephemeral form on computer screens or was hidden away on disks where no one could see it. For most people, it did not make sense to copyright something that was impossible to hold or see on a permanent basis.

That viewpoint began to change in the mid-1970s. The Commission on New Technological Uses of Copyrighted Works, established by the federal government in 1974, determined that computer programs constituted a creative work and therefore could be subject to copyright restrictions in the same way as literature, for example. In 1980, Congress revised the federal legal code to include computer programs in copyright legislation. Courts affirmed the new precepts in 1983, when the federal court of appeals for the Third Circuit ruled that Franklin Computer Corporation had violated copyright laws by copying code from Apple computers.[75]

In the face of these developments, it would have seemed pointless for GNU to attempt to reverse the legal power of copyright over software products and restore the days when computer programs did not fall under the purview of copyright law. However, Stallman and the GNU collaborators conceived an expedient that would allow them to turn copyright on its head and use it to prevent people from choosing *not* to share source code, rather than stifle copying.

The GNU project's attempt to use copyright in this way began in 1985, when Stallman released GNU Emacs with a copyright notice that granted users permission to make and distribute copies of the program and create modified versions of the software, so long as they did not claim sole ownership over the modified version. The Emacs copyright also required that all copies or derivative versions of the program carry the same licensing terms, which prevented other developers from distributing modified Emacs code under their own copyright terms.[76]

Stallman was not the first hacker to write a license for his software that guaranteed users the right to share code freely. In similar cases, however, the copyright notices amounted to brief, informal statements. For example, the 1985 version of the Unix program trn, a tool written by Larry Wall for navigating through lists of information, included a statement stipulating that "you may copy the trn kit in whole or in part as long as you don't try to make money off it, or pretend that you wrote it."[77]

Although the spirit of a license like Wall's was clear to hackers, the wording was ambiguous and potentially subject to challenges in court. If a programmer copied Wall's software, did he or she have to keep the same copyright license for the derivative work? Did trying to "make money" off of the code mean selling the software itself, or could it also include the sale of support or other services related to it? Were people who distributed the software obliged to state explicitly that Wall was its author, or would they remain compliant with the license so long as they did not expressly claim to have written the code themselves?

In contrast, Stallman's license for GNU Emacs was the sort of copyright notice that a lawyer could love. It clearly explained the terms under which users could share copies of the program with one another. It also specified how developers could legally modify or borrow the code to create their own software.

Other hackers in the GNU circle soon recognized the innovative power of the GNU Emacs license. By November 1986, they encouraged Stallman to broaden the license's language by replacing the word *Emacs* with *software* so that they could use the license to protect code they were writing for other programs. Stallman did so, turning the Emacs copyright notice into a "copyleft" license—a term Stallman borrowed from programmer Don Hopkins—that could apply to all software programs.[78] The new license became the first iteration of the GNU GPL.[79]

The GPL evolved from that initial version as Stallman and his fellow hackers worked to flesh out their copyright strategy fully. Between the first release of the GPL and the spring of 1988, GNU developers, perhaps recognizing that not all programmers wished to share code as freely as they did, introduced a number of changes to the licensing language that made it somewhat less restrictive. They specified that GPL-licensed software could be distributed alongside programs that were subject to other copyright notices. They required that developers commit to making the source code for GPL-licensed programs available only for a minimum of three years rather than in perpetuity. And they revised the GPL terms to permit programmers to distribute executables linked to system libraries without requiring them to distribute the source code for those libraries.[80] That last change was particularly important because it meant programmers could use the GPL to protect a program even if the program used software libraries (collections of functions that many applications on the same computer share in common) that were governed by other licenses.

GNU did not issue version 1.0 of the GPL, signaling that it deemed the license suitable for general use, until 1989. For the most part, version 1.0 contained the same terms as earlier iterations of the license. The major difference was that this "stable" version freed developers from having to share the modified version of a GPL-protected program with the community so long as they did not distribute the modified program publicly. Previous versions of the GPL had required developers to publish all code changes for a derivative work, even if they did not use the derivative work in a public setting or send binary copies to other users. This minor but significant change made it easier for programmers and organizations to adopt and modify GPL-protected software for private use, increasing the appeal of GNU programs and other products governed by the GPL. The 1989 GPL also contained language guaranteeing that future modifications of the GPL terms would not revoke the rights granted by an earlier version of the license.[81]

Licensing terms for software libraries remained a sticky issue for GNU following the release of GPL 1.0. As a result, the project in early 1991 announced an additional license, the Library General Public License, which later became the GNU Lesser General Public License (LGPL). Described by GNU as a "strategic retreat," the LGPL permitted developers to write programs that made use of GPL-protected software libraries even if the programs themselves did not use the GPL. That outcome was less preferable to requiring all software to be free, GNU declared, but it was a necessary pragmatic sacrifice because insisting that

GPL-licensed libraries "be used only in free software tended to discourage use of the libraries, rather than encourage free applications."[82]

Both the GPL and the LGPL have continued to evolve since the early 1990s, with GNU issuing two major version updates to the GPL in 1991 and 2007. They have proved hugely influential to the FOSS world since their introduction. Although the GPL is only one of more than one hundred free software licenses available today, data from recent years shows that a majority of FOSS projects have adopted some version of the GPL to protect their code.[83]

## THE BATTLE FOR "LOOK AND FEEL"

As innovative as GNU's copyleft strategy was, it did not prove to be a reliable remedy for the threats to free software that continued to evolve after the project found its footing. Chief among these challenges were the "look and feel" lawsuits and software patents that proliferated in the late 1980s.

The look and feel cases hinged on a shortcoming in software copyright. As the courts defined it in the 1970s, copyright could apply only to source code and machine code. This meant that developers could use copyright notices to prevent third parties from distributing binary copies of software or incorporating source code into another program without permission. But copyright could not stop someone from writing software that imitated the features or interface of another software product, as long as the imitation was based on original code.

Enter patents, which allow developers to protect the idea or concept behind a software program, as opposed to the code itself. Through patent claims, programmers could sue anyone who wrote code that emulated the way patent-protected software worked or looked—even if all of the code in the emulating product had been written from scratch and therefore violated no copyrights.

For GNU developers, patents posed a grave threat. After all, the project's main goal was to clone an operating system, Unix, over which GNU could claim no ownership. The possibility that AT&T might patent Unix's features—that is, the system's look and feel—threatened to undermine the entire GNU initiative.

Software companies' experiments with patent protections proved worrisome enough to GNU developers that, in June 1987, they publicly celebrated the legal victory of a company called SoftKlone over allegations that it had violated another organization's software patent by copying the look and feel of its programs. They did so even though SoftKlone developed proprietary software and was therefore antithetical to GNU's goals.[84]

The outcome of the SoftKlone case provided some assurance for GNU against the threat of software patents, but the project's leaders continued to fight them aggressively. In 1989, Stallman founded the League for Programming Freedom, which coordinated resistance against a series of lawsuits involving claims by Lotus Software that its competitors had improperly copied the features, but not the code, of its products. Stallman pitched the League as "an entirely separate organization" from the Free Software Foundation and GNU, adding that "the League for Programming Freedom is not an organization for free software, and it does not endorse the GNU project or the Free Software Foundation. Most League members write proprietary software, and some have founded companies that do so." In practice, GNU continued to promote the League's activities to its supporters throughout the late 1980s and early 1990s, reflecting the organization's deep and enduring concern over patents.[85]

GNU also endorsed a boycott of Apple from 1988 to 1995 because of the company's legal action against Hewlett-Packard and Microsoft regarding the look and feel of their software.[86] Calling the lawsuits an attempt by Apple to establish a monopoly in the personal computer market, GNU developers declared that "we will not provide any support for Apple machines."[87] Practically speaking, such actions had little effect on Apple, which eventually lost in court. (Even less potent was a campaign in 1988 by Stallman and some fellow hackers to distribute buttons bearing the inscription "Keep Your Lawyers Off My Computer" to combat Apple's actions.)[88] Still, the confrontation was notable for the precedent it set.

The relationship between Apple and FOSS advocates has been tense at many points in the past, not least because of Apple's incorporation of many free and open source programs into products that are otherwise highly proprietary, a trend chapter 4 discusses. Yet the antipathy was perhaps never as bitter as during the days when GNU urged its followers not to purchase Apple computers and refused to release software for them.

## HURD NOT SEEN

GNU achieved many impressive feats during the decade that followed Stallman's announcement of the project. But it failed to produce a viable version of the core piece of its operating system. That piece was the kernel—the program that binds the rest of a computer system together by allowing programs to communicate with hardware and one another.

By the beginning of 1990, all of the other main pieces of the GNU system were in place. The project had a debugger, a "fairly reliable" C compiler, an expansive library of C programming functions, and clones of all of the major Unix utilities. In a sign that the first step of Stallman's stated mission—creating a replacement for Unix itself—was nearly complete, the project by this time had begun developing other, nonessential applications for use with the system, such as the spreadsheet program Oleo ("better for you than the more expensive spreadsheet," GNU promised) and a tool for sending email. Developers who supported GNU were even adapting some GNU programs to run on MS-DOS, although GNU itself did not release software for that operating system.[89]

Yet despite GNU's impressive software catalog, work on the GNU kernel had not yet begun in a serious way. That was not by design. On the contrary, as noted above, when Stallman announced GNU in 1983, the first item on the list of programs that he said he intended to write was a kernel.[90] A series of changes and setbacks in GNU's plans for doing so chronically stalled the kernel development effort.

Stallman and the other early GNU collaborators at first planned to adapt a kernel called TRIX, which had been developed at MIT, for use as the core of GNU's Unix-like operating system. "It runs, and supports basic Unix compatibility, but needs a lot of new features," Stallman reported of TRIX in early 1986. He added that, most important, the TRIX kernel code was freely redistributable.[91]

In December 1986, GNU developers began work on a TRIX-based kernel implementation.[92] Within six months, however, they abandoned the effort and turned their attention toward another candidate, the MACH kernel. The June 1987 GNU newsletter reported that "We are negotiating with Prof. Rashid of Carnegie-Mellon University about working with them on the development of the MACH kernel. … If an agreement is reached, we will use MACH as the kernel of GNU."[93]

GNU developers had not entirely ruled out the adaptation of TRIX as their kernel at this time. They said they would return to that plan if they failed to secure permission to work with MACH. By this juncture, however, they had already lost much crucial time on kernel development, which had fallen far behind that of the other operating-system components they were building.

Little changed by January 1989. In that month, GNU reported that it still planned to use MACH as the basis for its kernel, but it had made no apparent progress toward implementing the software. Part of the delay stemmed from the fact that MACH incorporated some proprietary code from AT&T Unix, which MACH's original developers at Carnegie Mellon had promised to remove from the kernel.[94] While waiting for them to do so, GNU could reassure its supporters only that TRIX remained a possible back-up option in case the MACH plans did not pan out. It also mentioned a second potential alternative— the Sprite kernel that developers at Berkeley had created for use with BSD. By June, the plan evolved into using Sprite's file system in conjunction with MACH to build the GNU kernel.[95]

Carnegie Mellon programmers were slow in fulfilling their promise to free MACH of AT&T code. At the beginning of

1990, usage restrictions on MACH's code were proving a pervasive problem, and GNU reports on the kernel's status had grown more measured in their assessment of MACH's suitability for the project's goals. GNU informed supporters in January 1990 that its developers still "hope to use the Mach message-passing kernel being developed at CMU" and were awaiting the release of a free version of the kernel, which they expected by May 1990. The report went on to warn, however, that "until this happens, and we see precisely what is available and on what terms, we can't say for certain whether we can use it. We will not use Mach unless we can share it with everyone, and all users can redistribute it." The concern that distributing MACH outside of the United States could violate export laws had become another potential obstacle by this time, compounding the issues related to AT&T code.[96]

As the outlook for the MACH plan grew dimmer in 1990, TRIX and Berkeley Sprite remained alternative options for GNU. The enduring uncertainty over which code base GNU would end up adopting for its kernel, however, prevented meaningful development on any of the potential solutions.

The year 1991 finally brought definitive direction to GNU's kernel plans. Although a January 1991 report stated that GNU was still "waiting for CMU's lawyers to approve distribution conditions which will allow us to distribute the code" from MACH, in June the project announced that it had started development of a kernel based on MACH, which was by then free of AT&T code and legally exportable outside of the United States. "The system is intended to be both source and binary compatible with 4.4 BSD, and POSIX.1 compliant (when used in conjunction with the GNU C Library)," GNU informed is supporters.[97]

This meant that it would support the latest software programs designed to run on Unix-like operating systems.

GNU christened its MACH-based kernel *Hurd* because it was a collection of servers (or "herd of gnus," in Stallman's description) that ran on top of the MACH microkernel. Stallman originally intended to call the GNU kernel *Alix*, "after the woman who was my sweetheart at the time," but the lead kernel developer for GNU, Michael Bushnell, preferred the alternative term.[98]

In announcing the MACH-based Hurd kernel, GNU emphasized the technical advantages that its microkernel design promised. Most traditional Unix-like kernels adopt a monolithic kernel architecture. This means that all of the core services and functions—such as processing input from the keyboard and writing data to disk—that the operating system needs to provide exist in a single layer. In contrast, a microkernel delivers only a very basic set of services. It relies on separate programs, which are not part of the kernel itself, to do the rest of the work necessary to run the system. The computer science research community in the 1980s and early 1990s viewed microkernels as a promising innovation because their design could theoretically simplify operating systems by making them more modular and flexible.

Microkernels had their critics. Among them was Linus Torvalds, who, as the next chapter shows, decided to write Linux partially because he viewed microkernels as "essentially a dishonest approach aimed at receiving more dollars for research" by university programmers.[99] Yet GNU developers, who were intertwined with the academic community, espoused no such reservations. Thanks to Hurd's microkernel design, they promised that "a great number of functions, done in Unix by the kernel, will be done in the C library." That meant more flexibility in the way applications performed because programs would be less dependent on the kernel. Programs could "become much faster as well," GNU developers said.[100] They later went so far as to call the Hurd an entirely "new strategy of OS design," even though the project could not claim responsibility for having invented the microkernel concept.[101]

Despite such enthusiasm, Hurd development proceeded slowly. Developers made clear that GNU supporters should not expect the kernel to be complete for some time after they began work on it in 1991. In 1992, the project even suggested that it would adopt a different kernel as a temporary stand-in for the Hurd "to create an early, completely free GNU system" while Hurd matured.[102] A year later, GNU was calling for more volunteers to develop the Hurd, although the requirement that candidates should be able to "read and understand the source code with fewer than two questions, and have the time for a large project" did not help attract large numbers of programmers.[103]

Not until 1994 was the Hurd finally able to boot into a working system. A year later, in January 1995, it was capable of running most programs and was "right on the verge of being self-hosting (able to run on its own well enough to compile its own source code and be used for its own development)," according to GNU. "For a complete system we still have much more work to do," the project informed its supporters, but the developers envisioned issuing an alpha release in the near future.[104]

That release was slower in coming than the upbeat January 1995 report suggested. A year later, the Hurd remained "not yet ready for use."[105] Not until July 1996 did an alpha version, Hurd 0.0, finally appear and with it the first edition of the complete GNU operating system. The kernel at that point remained "very preliminary, and we don't recommend you try it unless you are in the mood to experiment," GNU cautioned. "Much work remains to be done on reliability, efficiency, and on user-level features to take advantage of the underlying capabilities." Yet the kernel developers were "making rapid progress on these tasks, and we plan to make further releases fairly often."[106] They produced a number of updated test releases of the Hurd during subsequent years.

Yet a stable, production-quality version of the Hurd has remained elusive up to the present. Although Hurd development continues today, the kernel does not support many modern hardware platforms. The most recent edition of the software is only version 0.6. In 2010, Stallman admitted that he was "not very optimistic about the GNU Hurd. It makes some progress, but to be really superior it would require solving a lot of deep problems."[107] Stallman has stated that GNU developers underestimated by about ten years the time that it would take to complete the Hurd kernel.[108]

In retrospect, the stunted development of the GNU kernel resulted, in part, from GNU developers' excessive emphasis on cutting-edge, untested ideas that theoretically could have made the Hurd better than other Unix-like kernels but that actually rendered development much more difficult than it needed to be. In March 1998, perhaps in a bid to convince programmers and users that the Hurd was better than the alternative Linux and BSD kernels, GNU pitched Hurd as a way for "users to create and share useful projects without knowing much about the internal workings of the system—projects that might never have been attempted without freely available source, a well-designed interface, and a multiple server [kernel] design."[109] Those were great goals, and they set the Hurd kernel apart from other Unix-like kernels. Without a kernel that actually worked well enough to use on production systems, however, GNU could not deliver the innovative features its kernel developers envisioned to real-world users.

Recurring attempts to rethink the Hurd design also negatively affected development. By 2000, in their quest for innovation, Hurd developers were wondering whether they should backtrack and port Hurd to work in conjunction with a micro-kernel other than MACH, which by that time was "no longer considered state of the art." Some members of the Hurd development team spent the next five years pouring their energies into the port, only to abandon the effort in 2005. This meant that more time was wasted in pursuit of technical sophistication at the expense of producing a usable kernel.[110]

It was ironic that unnecessary technical complexity ended up dooming the Hurd. Stallman told me in 2015 that he originally proposed the MACH-based Hurd as GNU's kernel because he believed it was a "purely pragmatic" way to build a kernel quickly. MACH seemed an advantageous choice because it "was a funded project at a university" and therefore posed little risk of disappearing or becoming something that GNU would have to develop without outside assistance.[111]

"I expected that using Mach, without having to write it, would save us a lot of the hard parts of writing a kernel," Stallman added. "I thought we could get it done fast and have the most elegant and powerful kernel."[112]

In reality, things "didn't work out that way; Mach didn't work so well, and eventually we had to maintain it ourselves," Stallman recalled. "Our design turned out to be a research project" rather than a pragmatic way to implement a working kernel.

Recognizing the extraordinarily complex approach that the GNU kernel team—of which Stallman was not a member—took to implementing the Hurd, he added that "perhaps the [Hurd] developer prioritized elegance over getting something out the door as fast as possible." Yet the main problem, he suggested, was simply that a number of design and implementation challenges that were hard to foresee in 1991 appeared as Hurd development proceeded.[113]

## GNU'S BALANCE SHEET

In practice, the Hurd's failure to materialize in usable form proved to be of little relevance for the success of either GNU or the broader FOSS world. As the next chapter explains, beginning in the early 1990s, the widespread adoption of the Linux kernel by the free software community bound together the rest of the GNU software and made it possible to run a complete system using only GPL-licensed software. This rendered the Hurd an obscure, mostly insignificant project by the end of that decade. Yet the setbacks that beset GNU kernel development throughout its history help to illuminate broader trends that explain what the project did well, what it could have done better, and how FOSS developers working on other software learned from GNU's mistakes.

In most respects, the GNU project was a runaway success. With the exception of a kernel, GNU produced highly stable replacements for all of the core software in Unix by the early 1990s. It went on to add many novel programs of its own, offering Unix users tools and features that they never saw on AT&T's operating system. GNU's programs were also better than the alternatives they supplanted. Studies in 1990 and 1995 found that GNU software was less than half as likely to crash as that of most commercial Unix systems of the time. In some cases, commercial Unix software crashed six times as often as GNU's.[114]

GNU's impressive success was the result of several factors. One was Stallman's intense and unwavering leadership of the project. As a man who was more willing than most people to sacrifice personal fortune and comfort in order to advance a project that promised no major material payoff, Stallman stood out even in an industry in which strong personalities are common. He gave up a job he loved to launch GNU from his Cambridge apartment. In contrast to most programmers who became titans of the IT world, Stallman never had a lucrative initial public offering or buyout to hope for when he took that gamble in 1983. (He received a $240,000 "genius" award from the MacArthur Foundation for his work with GNU and the League for Programming Freedom in 1990, but that was an unexpected, one-time windfall.)[115] The only reward he sought was saving hacker culture through free software.

It also mattered that Stallman stuck with GNU through its entire history. In other ventures within both the free and proprietary spheres of the software world, project leaders and executives commonly come and go for a variety of reasons, including political infighting, creative disagreements, and a desire for greater work-life balance. In this context, Stallman's long-standing commitment to GNU proved a boon for the project because it ensured constant and steady leadership. His sometimes abrasive personality and micromanaging tendencies were off-putting to some collaborators and users, as is discussed in later chapters. In other respects, however, Stallman became a model "benevolent dictator" (to borrow a term that FOSS programmers began using in the 1990s to refer to leaders of their community) for free software and the first of his kind to fill that role.[116]

GNU also owed much of its momentum to the community of enthusiastic users and programmers that organically arose around it. Most of the developers who participated in the project did not receive a salary from GNU. The few who did were paid only modestly, and unlike today, virtually no programmer during the 1980s or early 1990s who did not work for GNU or a company such as Cygnus could expect an employer to pay him or her to write free software.

Yet even without monetary lures, GNU was able to recruit highly capable hackers. At first, the project's volunteers, many of whom had worked at the AI Lab or were members of Stallman's personal circle, signed on because of Stallman himself. As GNU expanded, however, more developers began contributing because they embraced the software it was creating and the project's vision. According to Stallman, he "only knew the main ones" after GNU's roster of programmers became sizable, reflecting the growing size of the community he founded.[117]

At the same time, a strong community of users grew up around GNU. As this chapter has noted, early supporters of the project organized in locales ranging from Japan to Russia. Meanwhile, volunteers from across the United States who did not program helped to spread GNU software by hosting it on servers to which they had access, an important contribution at a time when sharing data over the Internet was much more difficult and expensive than it is today. Many more people supported GNU by simply using its software.

GNU's strong community played an important role in helping the project to acquire institutional support of various kinds, another key factor that explains much of its success. In addition to the cash donations, big and small, that sympathizers across the world started sending Stallman's way in the mid-1980s, gifts of office space, computer hardware, staff hours, and other resources from universities and companies sustained the project materially. GNU's leaders were careful never to associate themselves in a direct or official way with a particular organization apart from the Free Software Foundation. Nonetheless, they and the community they led adeptly built enduring relationships with independent institutions that proved to be vital for funding the project over the long term.

In forging these relationships, GNU benefited from the fact that entrepreneurs at companies like Cygnus had shown that it was possible to build successful businesses around free software. The development of commercial ventures related to free software helped GNU to position itself as more than a charity that gave away something. Instead, it generated products that were available at no cost and yet fed a thriving economic ecosystem.

The final and perhaps most crucial element to GNU's success was the recognition by Stallman and other hackers that creating free software required more than writing code. It also necessitated the production of auxiliary tools, chief among them licenses like the GPL, which protected free software against legal threats. In many respects, the GPL also became GNU's greatest success because the license went on to enjoy adoption by a wide variety of developers who never associated with the GNU project directly. The GPL even became the basis for licenses such as those of the Creative Commons organization, which, as chapter 6 explains, extended the concepts of the GPL software license to other realms of creative life.

The success of the GPL is also significant because it underlines the limits of GNU zealotry during the 1980s and the first years of the 1990s. This was another crucial, yet poorly appreciated, factor in GNU's success. As chapter 5 shows, portraying Stallman and his GNU stalwarts as uncompromising ideologues who alienated many potential developers and users was a common practice for advocates of the open source movement in the late 1990s. During the 1980s and early 1990s, however, when GNU was making its greatest strides in the realms of both software development and licensing strategy, the project and its leader remained more pragmatic than they have perhaps received credit for being.

The Free Software Foundation's willingness to scale back the restrictiveness of the GPL licensing terms as the license evolved prior to the GPL 1.0 release was one reflection of this pragmatism. Another was Stallman's support for Microsoft and HP in their case against Apple. Even if he staunchly disagreed with the way proprietary software companies distributed code, he was capable of making common cause with them when circumstances favored it. Third, GNU's cooperation with BSD developers, whose software did not at all meet the Free Software Foundation's definition of *freedom* because of its extremely liberal licensing terms, showed that Stallman through the early 1990s remained capable of compromise. He had not yet endorsed the Manichean thinking that, in later years, constricted the free software community's ability to reconcile with the open source camp.

GNU did well, but the project also made mistakes. It centralized development to too great an extent, a practice that helps explain why certain parts of GNU's software suite took a long time to build or never fully materialized. In contrast to projects such as Linux, GNU took little advantage of what today would be called a crowd-sourced approach to software development, especially in its early years. Rather than opening up development of all of its software to anyone who wished to participate and welcoming even small contributions, GNU assigned responsibility for the development of specific programs to particular individuals, who worked on them in isolation or in small teams. For this reason, the GNU development scene resembled what an observer such as Raymond would call the construction of a cathedral (a slow, tightly choreographed operation) much more than it did a bazaar (a loosely organized space that lacked centralized control).[118]

The cathedral approach seemed natural in the 1980s, when programmers adhered to the principles that Fred Brooks articulates in his influential 1975 book *The Mythical Man-Month and Other Essays on Software Engineering*.[119] Brooks argues that as more and more programmers work on a software project, bugs and communication difficulties rise at a rate that exceeds the additional productivity that a larger number of programmers contributes to a project. Comparing software production to gourmet cooking, Brooks suggests that good software development takes time and proceeds most efficiently when programmers work independently or in small groups on discrete projects of limited scope, which later can be integrated to produce a large system like GNU's software suite. Programmers came to know this precept as Brooks's law.

For most of GNU's programs, Brooks's approach to development worked well enough. It became a problem when the individual assigned to a particular project under GNU's direction failed to deliver, creating a single point of failure. This was the case for the gsh GNU shell, for example. GNU's Brooks-style programming strategy also arguably bogged down Hurd development. The fixation of Hurd's small group of programmers on an overly complex kernel architecture might have been redirected if a greater diversity of minds had contributed to the kernel project and inclined it in a more pragmatic direction.

By the late 1990s, after the Linux group had shown that the bazaar mode of development could work and widespread Internet connectivity made it easier for developers in remote locations to collaborate, GNU began to experiment more extensively with crowd-sourced development. In 1997, for instance, the project announced that "volunteers with a PC are … eagerly sought" to help test the 0.2 release of the Hurd and identify bugs.[120] GNU developers were attempting to imitate what Linux developers had already been doing for years by that time—that is, rely on users to debug software rather than do it themselves. Because GNU did not adopt such development strategies earlier, it missed an opportunity for innovation, leaving it to Torvalds and his followers to pursue the most enduring experiment with crowd-sourced software production over the Internet.

GNU's second big mistake was trying to find too many shortcuts to producing free software. Stallman and his collaborators lost valuable time attempting to borrow code from other programmers in order to jump-start the development of GNU programs like the compilers and the kernel. Ultimately, after discovering that free code that met their needs was in short supply, they ended up writing most of these programs themselves.

It is hard to fault GNU for looking for shortcuts. One of the leading arguments in favor of FOSS today is the utilitarian suggestion that it saves programmers time and energy because sharing code obviates the need for developers to reinvent the wheel whenever they begin working on a program. In experimenting with third-party programs such as VUCK and Gosmacs or hoping that code from the TRIX or MACH kernels would suit their needs, Stallman and his followers were simply trying to save valuable time and effort so that they could invest them in other GNU endeavors.

GNU planners perhaps did not always know when to cut their losses, however, and to begin writing their own code when it became clear that borrowing from other programs was unlikely to work. In addition, attempts to adapt third-party code restricted developers' ability to design the GNU system in the ways that they deemed best. Instead, they had to adapt many plans to fit the requirements of software that originally had been written elsewhere, for other purposes. That limitation constricted GNU's capacity to innovate in some respects.

Lastly, GNU suffered from short-sightedness regarding the importance of personal computers and the market that grew up around them. In 1986, Stallman expressly rejected the notion that GNU should build software that ran on microcomputers or should create a clone of an operating system designed for them. "Why not imitate MSDOS or CPM?" he rhetorically asked GNU supporters, referring to the leading operating systems for PCs of the time. He answered, "They are more widely used, true, but they are also very weak systems, designed for tiny machines." In lieu of catering to the microcomputer crowd, GNU invested its resources in developing a clone of Unix that could run on the large computers owned by universities and other institutions. Stallman deemed this type of software "much more powerful and interesting" than anything designed for PCs.[121]

This remained GNU's attitude through 1992, when the project cautioned users who were curious about running its software on PCs: "We do not provide support for GNU software on microcomputers because it is peripheral to the GNU Project." By this time, other developers, who had no direct affiliation with GNU, were working on GNU software for PCs, and GNU was willing to share information about their work, but it declined to engage in development for microcomputers itself.[122]

Not until 1993 did GNU do an about-face by distributing ports of its software for MS-DOS environments.[123] That decision placed GNU directly in the PC world, which it embraced with fervor. By this time, however, the project was late to the party. Other free, Unix-like operating systems (including BSD/386 and GNU/Linux distributions) that other people had created were already running on PC hardware. Programmers who were not affiliated with GNU could take credit for having ported many of GNU's programs to work on PCs.

It would be unfair to criticize Stallman for his decision in the mid-1980s to create a free version of Unix rather than something that catered to the PC community. At the time, Unix and institutional computers from companies such as Digital Equipment Corporation (DEC) remained the platforms on which the greatest innovations were taking place within research communities. They were also the environments hackers knew and loved. It took several years before falling prices for PC hardware, the proliferation of affordable software programs for PCs, and the growing use of the Internet by individuals outside the research and programming communities brought these machines to the fore of the computer world.

Nonetheless, one wonders what would have happened if Stallman had set out in 1984 to create a free replacement for MS-DOS rather than Unix. The system he would have built would almost certainly not have been as powerful, from the perspective of developers, as the one GNU actually became. And Bill Gates's marketing acumen may well have thwarted its success. But if a good, free PC operating system had existed in the 1980s and PC sellers and users had adopted it for use with their systems, the monopolistic dominance that Microsoft later established over much of the computer industry may well have been nipped in the bud.

In the long run, the software that GNU developers produced helped to form the basis for systems that eventually became a powerful alternative to Microsoft's products. But a key part of that story involved the initiative not of GNU but of the young, irreverent programmer from Finland who created the Linux kernel. The next chapter tells his story.

# 3 A KERNEL OF HOPE
## The Story of Linux

### THE FALTERING REVOLUTION:
### GNU AND BSD IN THE EARLY 1990S

FOR POLITICAL REVOLUTIONARIES in eastern Europe and central Asia, 1991 was a good year. By the close of that year, the Soviet Union had collapsed, leaving in its wake fifteen independent states and fledgling democracies.

In contrast, the GNU free software revolutionaries had much less to celebrate by the end of 1991. It appeared increasingly uncertain that the work they had pursued over the previous seven years would bring the FOSS revolution to fruition and make the world safe for hackers once again.

True, GNU had accomplished much by this time. As the previous chapter notes, the project had released feature-rich versions of almost all of the programs required to build a free, Unix-like operating system. People around the world were using GNU software. GNU developers were amassing cash donations in the six-figure range and enjoying strong support from a variety of companies and universities.

Yet GNU was stumbling in certain key respects. One of these was in the realm of kernel development. As the previous chapter explains, GNU programmers did not finally start writing the Hurd kernel until the middle of 1991, and things did not go well from there. Without a kernel, the GNU system that Stallman had envisioned in 1983 remained akin to a mansion without a roof or a jet fighter without an engine: it was sophisticated and complete in all respects except for the one that mattered most.

Kernel problems were only one of the serious challenges GNU faced in the early 1990s. The others (also detailed in the previous chapter) involved GNU's failure to take PC hardware and software seriously until well after they had assumed an outsize role in the computer market. The centralized development approach that Stallman and his collaborators clung to also was a problem. The latter issue proved particularly detrimental for GNU because it meant that, for programmers, writing GNU code entailed basically the same type of work as coding for a commercial software company. From a developer's standpoint, there was nothing different about the way GNU created software that set the project apart from other major development initiatives. The GNU code was free, and that mattered to a lot of hackers. But coding for GNU was no more fun than coding for any other project. That fact helped to create an opening for Linux, which offered developers a fundamentally new model of collaboration, free of centralized hierarchies and lethargic release schedules.

As a result of these setbacks, by 1993 people began asking (to quote the title of a *Wired* magazine article) "Is Stallman Stalled?"[1] GNU had grown "bogged down," observers said, and GNU developers were sensing that their "window of opportunity to introduce a new operating system" was quickly passing by, if it had not already disappeared—as Robert Chassell, one of the founders of the Free Software Foundation, put it at the time.[2]

Meanwhile, hackers had reason to doubt the prospects of the free software community's other great hope for building a free implementation of Unix. The BSD NET 2 operating system from Berkeley was released in 1991, as the previous chapter explains, and several derivative ports for PCs appeared during the two years that followed.[3] For a time, BSD seemed promising to hackers as an alternative to AT&T Unix. Even the GNU programmers, despite viewing the BSD licensing terms as insufficient for building a free operating system, saw enough value in NET 2 to start distributing copies of it by the summer of 1992 as part of the Free Software Foundation's software-distribution service.[4]

Yet the early momentum that the BSD derivatives enjoyed did not endure. For two main reasons, none of the BSD-based operating systems proved to be a good substitute for the type of system that hackers like Stallman envisioned. The first problem stemmed from the licensing terms under which the BSD software was released. As noted above, the Berkeley license that governed the BSD derivatives did not require people who distributed the software to provide source code with it. This meant that the BSD licenses failed to protect the hacker imperatives of transparency and sharing.

The second, more significant issue affecting the BSD derivatives involved legal troubles. In January 1992, a company called Berkeley Software Design, Inc. (BSDI) began selling a commercially supported version of the BSD NET 2 operating system with a price tag of $995—which, although steep for an individual consumer, was 99 percent less expensive than AT&T's commercial Unix, as BSDI's advertisements noted prominently.[5] Soon after the platform hit the market, however, Unix Systems Labs, which owned the Unix trademark at the time, sued BSDI, alleging that the company infringed its trademark by describing BSD-based software as a form of Unix. It also charged that NET 2 contained copyrighted Unix code.

The case was settled out of court the next year. But the Regents of the University of California then countersued Unix Systems Labs. The university contended that the company had not properly acknowledged the BSD code that formed part of the System V version of Unix, as was required by the license agreement under which the University of California had released BSD.

The parties moved toward settlement in June 1993, when Novell acquired Unix Systems Labs. Novell's CEO at the time, Ray Noorda, began talks with Berkeley at the end of the summer, and the parties reached a settlement the following February.[6] As a result of the lawsuit, only three files were removed from the eighteen thousand that comprised the NET 2 code, and minor changes were made to some others. The legal challenges that BSD faced had no meaningful effect on the ability of hackers or anyone else to use and redistribute BSD-based systems freely.

Yet the challenges did serious damage. During the nearly two years between the start of the Unix Systems Labs lawsuit and the final settlement, uncertainty over whether users of BSD-based operating systems might be compelled to purchase an expensive license to run versions of BSD stymied adoption of the systems. Also problematic was the settlement's requirement that BSD developers rewrite some minor parts of the BSD code that the agreement deemed to be in violation of the Unix Systems Labs copyright. The changes were not extensive and entailed relatively little effort. But they created a distraction for BSD programmers at a crucial moment in the system's development, when they might otherwise have been able to focus on implementing novel features instead of rewriting functions that already existed.[7]

The free versions of BSD did not go extinct in the early 1990s, and many survive into the present, as the next chapter notes. However, Berkeley's decision to disband the Computer Science Research Group, which had been the center for BSD development, following the release of the final version of BSD in June 1995 meant that the BSD community lost its central reference point. The various BSD derivatives that remained diverged in certain respects from one another. It is undeniable that the BSD legacy remains central to the FOSS world today: as many as half of the utilities in most GNU/Linux distributions descend from BSD code.[8] Nonetheless, because of both the Unix Systems Labs lawsuit and the BSD licensing terms, BSD-based operating systems never evolved into a satisfying solution for most hackers seeking a Unix-like operating system that promised to be free as in freedom.

## THE MAN BEHIND THE KERNEL

While GNU listed and BSD NET 2 faced a stillbirth, a new generation of hackers dreamed of an operating system that would combine the allure of a free version of Unix with the accessibility of PC hardware. It took one of the generation's own members, Linus Torvalds, to deliver on that vision.

In many ways, it was no surprise that Torvalds ended up producing the kernel that eluded teams of seasoned Unix hackers at both Berkeley and GNU. Geographically as well as technologically, Torvalds grew up in a very different world than that of the generation of programmers who preceded him. His background helped him to think in new ways about old programming problems.

One important element of Torvalds's experience was that, in contrast to programmers like Stallman or the BSD developers at Berkeley, he had only minimal exposure to advanced academic computer science research when he began working on the Linux kernel. He eventually completed a master's degree in computer science but not until 1996. When he introduced Linux to the world in the summer of 1991, he was not even close to finishing his undergraduate degree.

The difference in this regard between Torvalds and Stallman, who had nearly a decade of experience under his belt working as a programmer at one of the world's leading technical universities when he founded GNU, is striking. It also suggests part of the reason that Torvalds readily embraced a decentralized, Internet-based approach for developing Linux—even though such a strategy would have seemed anathema to most professional programmers at the time, weaned as they were on Brooks's law.

Another key distinction for Torvalds was that most of his early experiences with computers involved PCs. As a teenager, Torvalds used a Commodore VIC-20 machine, predecessor to the better known Commodore 64, that he inherited from his grandfather.[9] Around his eighteenth birthday, using money he earned cleaning city parks in Helsinki, he replaced the Commodore with a QL computer from a British company named Sinclair.[10] Torvalds's third computer, which he purchased as an undergraduate student on the eve of beginning the programming venture that resulted in Linux, was a PC with an Intel 386 processor.[11] These machines did not look or work anything like the powerful, expensive PDP and VAX computers that the Unix hackers of old had learned to program on—and that developers like Stallman continued to regard as the only hardware platforms worthy of hosting an operating system as sophisticated as GNU's.

Torvalds did not encounter a large computer running Unix until the fall of 1990, when the University of Helsinki, where he was a student, acquired a MicroVAX computer that sported Ultrix, Digital Equipment Corporation's implementation of Unix.[12] By then, the young hacker had acquainted himself with operating systems that mimicked Unix but ran on the PC hardware with which he had grown up. To Torvalds, building a Unix-like operating system for the PC seemed like the obvious thing to do from the start, a characteristic that distinguished him in a crucial way from the hackers of Stallman's generation.

It also mattered that Torvalds lived in Finland, far from the major centers of computer science research at the time. Torvalds's hometown, Helsinki, had a university with a strong computer science department and as much Internet connectivity as most other major European cities of the time. Indeed, Finland was an early hub of computer research in northern Europe.[13] Yet because the country's computer industry remained minuscule compared to that of larger nations, Torvalds often had to order computer parts and programs from abroad, especially during the years when his main computer was the British Sinclair. International snail-mail delivery of such items could take months. That was perhaps one factor that attracted Torvalds to distributing software via the Internet, which became crucial to the way he developed and shared Linux. The fact that he could meet relatively few computer science experts in Helsinki meant that the Internet also served as his lifeline for communicating with developers and exchanging ideas, as he did when announcing Linux over Usenet. In these respects, Torvalds's experience was distinct from that of the GNU and BSD programmers operating out of Cambridge and Berkeley, where they were at the centers of their respective hacker universes. In contrast, Torvalds was in a remote galaxy.

Just as Torvalds grew up on the geographic margins of the hacker community that he later helped to lead, he also belonged, in certain ways, to a marginal community within his own society. He was born into Finland's tiny Swedish-speaking minority, which constituted around 6 percent of the country's population at the time of his birth. Swedish-speaking Finns have long lived happily among their neighbors, and there is no evidence that Torvalds's linguistic background proved a major obstacle for him. (He also became fluent in English, the *lingua franca* of programmers on the Internet, at an early age.)[14] On the contrary, it placed him in a community that was at the forefront of the computing industry in Finland, where a majority of managers at companies such as IBM Finland belonged to the Swedish-speaking minority.[15]

Still, Torvalds's experience living outside the mainstream Finnish population may have helped him to think differently about projects like Linux. Torvalds lived on the periphery of the dominant linguistic community in his country, just as the kernel he developed existed on the margins of the mainstream software culture of the early 1990s, defying the norms of both proprietary and free software programmers alike.

Finally, Torvalds has stated that his family had an important effect on some of the decisions he later made regarding Linux. "I undoubtedly would have approached the whole no-money thing a lot differently if I had not been brought up under the influence of a diehard academic grandfather and a diehard communist father," he wrote in his 2001 autobiography to explain why he was committed to releasing Linux free of charge.[16] His grandfather, one of Finland's first professional statisticians, died in 1983, when Torvalds was in his early teens. And Torvalds's father eventually moderated his hardline communist beliefs. Nonetheless, these two patriarchs imbued Torvalds at a young age with an appreciation for the value of research and exploration and radical notions of sharing and anticapitalist mores, and they shaped how Torvalds thought about the best ways to design and distribute software.

All of these experiences help to explain why Torvalds produced a novel operating system kernel and pioneered a radically new mode of software development that varied in crucial ways from what older hackers in the United States were building. Yet the specific event that set him on the path toward developing Linux was his purchase in the summer of 1990 of the book *Operating Systems: Design and Implementation*, which Andrew S. Tanenbaum, an American computer science professor teaching in the Netherlands, had published in 1987.[17] According to Torvalds, the book, which exposed him to "the philosophy behind Unix and what the powerful, clean, beautiful operating system would be capable of doing," changed his life and "launched me to new heights."[18]

"As I read and started to understand Unix," he added, "I got a big enthusiastic jolt. Frankly, it's never subsided."[19]

## WHY LINUX?

Tanenbaum's book was not about AT&T's version of Unix, though. It focused on a Unix-like operating system called Minix, which Tanenbaum created after receiving a Ph.D. from Berkeley and working for some time with the Unix group at Bell Labs.

According to Tanenbaum, his chief goal in writing Minix, which he coded from scratch entirely on his own, was to provide his computer science students with a Unix-like operating system whose source code was much less expensive than Unix's code. But it gained a broader following. Within a couple of months of Tanenbaum's release of the first version of Minix in 1987, the system "became something of a cult item, with its own USENET newsgroup, comp.os.minix, with 40,000 subscribers," he wrote. "Many people added new utility programs and improved the kernel in numerous ways."[20]

Tanenbaum, however, was reluctant to extend Minix—which was so-named because it was a miniature version of Unix—beyond the basic teaching tool he had designed it to be. In his recollection, "I didn't want it to get so complicated that it would become useless for my purpose—namely, teaching it to students." He added that there was every reason to believe that when Minix appeared, GNU and BSD were on the verge of providing a free, production-quality Unix implementation that would far surpass anything Tanenbaum could develop on his own.[21]

From the perspective of someone like Torvalds, however, who began using Minix shortly after acquiring an Intel-based PC in early 1991, the chief importance of Tanenbaum's operating system was not its use for teaching.[22] It was that Minix was one of the first Unix-like operating systems that had been designed to run on microcomputer hardware rather than the PDP and VAX machines on which GNU and BSD development centered. In addition, although Minix cost some money (which irked Torvalds, as noted below), it was a teaching tool rather than an explicitly commercial product. That set it apart from other Unix-like operating systems for microcomputers, such as Coherent and Xenix.

Yet for Torvalds, Minix was not at all an adequate PC-based Unix implementation. In his view, Tanenbaum's lack of interest in extending the platform to include more features represented a major shortcoming. Torvalds was particularly disappointed that Minix did not work well on his PC, which had an Intel 386 processor. Minix ran on this type of computer, but because Tanenbaum had designed the operating system for a different family of microcomputers, it supported 386 chips only with the assistance of a special patch written by Bruce Evans. The patch was difficult to install, and after it was applied, it left much to be desired for someone like Torvalds who wanted to get the most out of a Unix-like environment on a 386 PC.

Minix also lacked terminal emulation, a feature that made it possible to log in to remote computers. That deficiency prevented Torvalds from using Minix on his home PC to connect to the Unix computer at the university where he was a student.[23] Minix "had been crippled on purpose, in bad ways," because Tanenbaum "wanted to keep the operating system as a teaching aid," according to Torvalds.[24]

Another shortcoming of Minix, in Torvalds's view, was that it used a microkernel. As the previous chapter noted, Torvalds has strongly criticized microkernel architectures on technical grounds. Although his opinions on that topic may not have been fully developed before he started writing his own kernel, in early 1992 Torvalds felt strongly enough that Minix was poorly designed to complain to Tanenbaum that it was filled with "brain-damages," which derived from its microkernel architecture.[25] Torvalds also criticized Minix for its lack of portability and compliance with the POSIX standards of Unix-like operating system design.[26]

But the technical features that Minix lacked were only part of its problem for Torvalds. The licensing parameters that Tanenbaum imposed, particularly those that required payment of a fee to run Minix, seemed even worse. "Look at who makes money off minix, and who gives linux out for free," Torvalds seethed in a January 1992 Usenet post, in which he excoriated Tanenbaum for charging for Minix: "Then talk about hobbies. Make minix freely available, and one of my biggest gripes with it will disappear."[27] Torvalds found the cost of Minix—which amounted to $169 "plus conversion factor, plus whatever" someone in Finland might have to pay to acquire a Minix license—"outrageous at the time," he wrote in 2001, adding, "frankly, I still do."[28] He told me in 2016 that "free as in 'gratis' was actually an earlier concern than the whole 'free as in freedom'" consideration in motivating his decision to write Linux.[29]

Tanenbaum took offense at the criticism of Minix's cost, particularly because he thought he was doing cash-strapped students like Torvalds a favor by releasing Minix as a low-cost way to study Unix system design. In the same Usenet discussion mentioned above, Tanenbaum told Torvalds that, in writing Minix, "an explicit design goal was to make it run on cheap hardware so students could afford it." Referring to the GNU project, which was producing free software that for the most part ran only on very expensive computers designed for purchase by institutions, Tanenbaum sarcastically added, "Making software free, but only for folks with enough money to buy first class hardware, is an interesting concept."[30]

Tanenbaum has stuck by this argument. In an essay on Minix's history that he posted on the Internet in the early 2000s, he noted that Minix was never "free software in the sense of 'free beer'" (meaning that it was never free of cost) but that its source code was always freely available. In addition, the cost of a Minix license was a tiny fraction of what users paid for a commercial version of Unix. "By 1987," he explained, "a university educational license for UNIX cost $300, a commercial license for a university cost $28,000, and a commercial license for a company cost a lot more. For the first time, MINIX brought the cost of 'UNIX-like' source code down to something a student could afford."[31]

Yet Torvalds's abiding irritation with the fact that Minix and similar operating systems cost anything—even if Minix was much more affordable than the alternatives—reveals much about how thinking about money affected the nascent Linux community. Because Linux and the GNU software that accompanied it in the late 1990s became the centerpiece of a flourishing commercial ecosystem with Torvalds's blessing, it has been easy for FOSS users to forget how opposed Torvalds was to the notion of making money off of Linux early on and how important that factor was in pushing him to start writing the kernel. In fact, Torvalds's animosity toward the prospect of profiting from software represented a crucial dimension of why he chose to write Linux, and it made his work different from that of Stallman and GNU.

Torvalds's choices are not fully explained by any of the other factors that influenced him. For one, there is little evidence that the philosophy of the Free Software Foundation had much effect on Torvalds when he crafted Linux. Torvalds attended a speech that Stallman gave "probably in 1991 or so," in Torvalds's recollection, at the Helsinki University of Technology, which introduced him to the Free Software Foundation's ideology.[32] That may have been part of the reason that he complained in October 1991 about Unix-like operating systems that "come with no source" and as a result "are ideal for actually using your computer, but if you want to learn how they work, you are f--ked."[33] That remark made it clear that Torvalds cared about keeping source code available for the utilitarian purpose of ensuring that other programmers could understand how it operated.

That sentiment, however, was different from sharing source code as a matter of moral principle, as Stallman did. Torvalds wrote that in 1991 he "wasn't much aware of the sociopolitical issues that were—and are—so dear to [Stallman]. I was not really all that aware of the Free Software Foundation. ... judging from the fact that I don't remember much about the talk back in 1991, it probably didn't make a huge impact on my life at that point. I was interested in the technology, not the politics."[34]

The deficiencies that Torvalds perceived in Minix also do not account, on their own, for his decision to produce his own kernel. If those had been the only issue at play, Torvalds could simply have written a terminal emulator for Minix, tweak the Minix code so it would run better on Intel 386 hardware, and left it at that. Torvalds may have disagreed with the microkernel design of Minix, but that did not significantly affect his ability to use the operating system.

Neither was the joy of coding the main factor in Torvalds's decision to write Linux. It is true that he described programming in his autobiography—which was titled *Just for Fun*, implying that having fun was a large part of the reason for writing Linux—as "the most interesting thing in the world. It's a game much more involved than chess, a game where you can make up your own rules and where the end result is whatever you can make of it."[35] Depicting oneself as a developer who was passionate about coding for coding's sake might seem to be a mere public relations move. But there is good reason to believe that Torvalds was expressing genuine sentiment through such statements. The time and effort he invested into writing the first versions of Linux were considerable, and it is difficult to imagine a rationale human being undertaking such an endeavor, especially with no strong financial incentive, unless that person truly and deeply enjoyed the art of programming.

Yet personal amusement does not explain why Torvalds chose to write Linux or give it away for free to other people. If he just wanted to have fun programming, he could have written plenty of other programs rather than one that duplicated much of the functionality already available to him from Minix. And if Linux was purely about having fun for Torvalds, there was no reason for him to invite others to contribute to his kernel, which reduced the amount of fun-inducing programming he had to do for himself.

So if free software ideology, shortcomings in Minix, and a desire to have fun do not fully explain why Torvalds wrote Linux, what does? The answer centers on money. Torvalds wanted a Unix-like kernel that was totally free of cost. As noted above, his chief complaint about Tanenbaum's Minix system in the early 1990s was its price tag. That issue remained a major point of contention for Torvalds at least as late as the 2001 publication of his autobiography—even though, by that time, he was a millionaire with plenty of cash at his disposal. As recently as 2013, Torvalds declared, with his characteristic humor, that "software is like sex: it's better when it's free."[36]

To note that the most significant motivation for Torvalds in writing Linux was to obtain an operating system that cost no money is not to dismiss the authenticity of his intentions or the validity of his work. On the contrary, as Tanenbaum himself suggested when contending that GNU's designers failed to give enough weight to hardware cost when choosing which platforms to support, there was plenty of good reason in the early 1990s to design a system that cost absolutely no money and ran on affordable PCs.

Yet the opposition to profit does mean that Torvalds thought very differently from the GNU developers in one key respect and that his work appealed primarily to a different set of users. Although the GNU project never required users to pay for its software if they copied it among themselves or downloaded it from the Internet, the project charged significant fees—up to $5,000—for official distributions of the code on disk as a way to support itself, as the previous chapter notes. GNU was not concerned with making software free of cost for anyone, even those without the ability to download or copy code. If it had been, the project would likely have found a way to distribute its software on disk for free (as the organizers of the Ubuntu GNU/Linux distribution did in the 2000s via the ShipIt program, for instance).

For all of these reasons, the fact that no fully functional Unix-like operating system was available from GNU or elsewhere free of cost in 1991 gives the key to Torvalds's decision to write Linux. It also explains why his August 1991 Usenet post announcing Linux to the Minix community, which is quoted below, notes prominently that the operating system was "free" (with the implication that it cost no money) yet makes no mention of whether Torvalds intended to allow other people to view and modify his source code freely—which underscores how unimportant the Free Software Foundation's philosophical message was to Torvalds.[37] And it accounts for his decision, detailed below, to release the first versions of Linux under a license that prevented anyone from using the code to make money.[38] Only later did Torvalds's thinking about the relationship between software and money evolve in a way that led him to adopt the GPL license for the kernel, making it possible for Linux and GNU software to join forces to advance the FOSS revolution— and for people to make money using Linux code.

## FROM MINIX TO LINUX

Although Torvalds's greatest complaint about Minix was its cost, his desire to shore up some of what he viewed as the operating system's technical deficiencies sparked the beginnings of Linux—even though designing a complete kernel was not on his agenda when he first went to work.

The first release of the Linux kernel started as an effort by Torvalds to write a terminal emulator for Minix. As noted above, the absence of this type of program in Tanenbaum's operating system prevented Torvalds from using Minix on his home computer to connect remotely to his university's Unix server. To produce his own terminal emulator, Torvalds had to delve into the lower-level functionality of the 386 computer chip. That endeavor entailed tedious work. But for Torvalds it was a bonus because it meant that creating the terminal emulator would be a way to explore the intricacies of the computer hardware he owned. Toward that end, he wrote the emulator in assembly language, which was much more complex than a higher-level programming language, "just to learn about the CPU."[39]

Torvalds's terminal emulator for Minix was complete by the spring of 1991, and he could use it to log into his university's Unix system. Yet he was not satisfied. The emulator in its first iteration lacked disk and file system drivers, and without them, it was impossible to upload or download files from the remote server. So Torvalds went to work adding those features to his program.

The ground that Torvalds needed to cover to extend his terminal emulator into a complete operating-system kernel was narrowing because reading and writing to disks is one of the core responsibilities of a kernel. By the time he started work on the disk and file system drivers, "it was clear the project was on its way to becoming an operating system" rather than just a terminal emulator, Torvalds recalled.[40]

Perhaps because he trod a path similar to Stallman's, Torvalds referred to his project early on as the "gnu-emacs of terminal emulation programs."[41] As the previous chapter notes, Stallman's endeavor to build the GNU operating system started with the GNU Emacs text editor, a far cry from a complete operating system. For Stallman and GNU, Emacs was hardly the most essential part of the system, nor was it the most difficult to write. But it was the seed from which the rest of the GNU system sprang, at least conceptually. Torvalds's Minix terminal emulator, which was a similarly nonessential part of the kernel he eventually produced, did the same thing for Linux.

In descriptions of Linux's early history—or what could be called its prehistory because this was the period before Torvalds had fully conceived the kernel or even given it a name—Torvalds is unsure exactly when he made the choice to extend the terminal emulator into a kernel. He has written only that this happened sometime around April 1991.[42] Perhaps the decision evolved too gradually to associate with a particular date, or maybe Torvalds never consciously made it until after he was already well on his way to writing a kernel.[43]

What is certain is that Torvalds had a kernel on the brain by July 1991. In that month, he posted to the comp.os.minix Usenet newsgroup requesting links to a current version of the POSIX standards definition from the IEEE (Institute of Electrical and Electronic Engineers) Computer Society.[44] POSIX, an acronym for Portable Operating System Interface (with an X added in traditional Unix fashion), specifies how a Unix-like operating system should be designed to be feasible for porting to different hardware platforms. The standard also ensures compatibility with any software applications written for POSIX compliance. By asking about the POSIX standards, Torvalds made clear that he was at least considering writing a kernel that would serve not only his personal purposes but also those of others interested in a new Unix-like operating system.

No one took serious notice of Torvalds's request at the time, and no Usenet readers responded with tips about obtaining the POSIX standards definition. As a result, Torvalds mined the information he sought from manuals at his university that described Sun's implementation of Unix. He also referred to Tanenbaum's book about operating-system design. Neither source provided the complete reference on POSIX standards that Torvalds sought, but they sufficed to allow him to continue to work on what was becoming the Linux kernel.[45]

Progress was slow at first, and this was not only because Torvalds lacked access to full POSIX documentation or prior experience programming something as complex as a kernel. His enthusiasm made up for those obstacles, but it did not make it any easier to contend with the complexities of the Intel 386 computer-chip architecture. Nor did it resolve the difficulty of debugging a kernel system during the early stages of development, when the code remains so incomplete that it is not possible to take advantage of sophisticated debugging tools, which help programmers identify the parts of code that cause a program to crash.

Asked in 1992 how he debugged early versions of the kernel, Torvalds wrote that what made the work so tedious was that he could not "even think about debuggers" at the time because none of those available to him worked well with Intel 386 hardware.[46] His code remained so basic that even printing error messages on the screen when a crash occurred was not possible.

Lacking sophisticated debugging resources, Torvalds explained, he adopted an approach that was pragmatic yet crude from the perspective of programmers like those at GNU, who wrote elegant, sophisticated code and had advanced debugging software at their disposal:

> What I used was a simple killing-loop: I put in statements like `die: jmp die` at strategic places. If it locked up, you were ok; if it rebooted, you knew at least it happened before the die-loop.[47]

In less technical terms, Torvalds meant that he inserted snippets of code at various places in his kernel program that he called *die-loops*, which caused the kernel to cease executing and freeze. If the system rebooted—as it did whenever the kernel crashed due to a code flaw—he would know that the issue had occurred at a point in the program prior to where he inserted the die-loop. If the system merely froze without rebooting, Torvalds knew that his program had run successfully up to the point in the code where the die-loop appeared. This approach to debugging was tedious and primitive, but it worked.

Progress became smoother after Torvalds had implemented enough of the kernel to "have a minimal system up [that] can use the screen for output." Yet even then he had to resort at times to the ugly die-loop approach described above. "All in all, it took about 2 months for me to get all the 386 things pretty well sorted out," to the point necessary to use more sophisticated debugging tools, Torvalds wrote in an early account of Linux's history.[48]

With the kernel code stable enough to accommodate more complicated debugging strategies, Torvalds implemented task switching, a key component of a Unix-like kernel, which allows users to switch between different applications. He next wrote keyboard and serial drivers so that his kernel could communicate with the input devices connected to the computer. With those components in place, Torvalds finally began to enjoy "plain sailing: hairy coding still, but I had some devices, and debugging was easier. I started using C at this stage" rather than assembly language, a change that "certainly spe[d] up development. This is also when I start[ed] to get serious about my megalomaniac ideas to make 'a better minix tha[n] minix.'"[49]

Lack of reliable documentation on the finer points of Intel's 386 hardware remained a persistent problem for Torvalds. But he eventually overcame the challenge and completed a file system, the last part of the code that was required to have a basic equivalent of the Unix kernel. Torvalds's kernel at that point "wasn't pretty, it had no floppy driver, and it couldn't do much of anything. I don't think anybody ever compiled that version," he recalled in 1992.[50] Nonetheless, "by then I was hooked, and didn't want to stop until I could chuck out minix" completely, replacing it with the kernel he had built on his own.[51]

It was at this point, in August 1991, that Torvalds announced his kernel project to the world in a Usenet post on comp.os.minix that has since become something of a legend among Linux acolytes. Confirming the suspicions of users who had read his query about POSIX documentation a month earlier, Torvalds opened a Usenet thread on August 25, 1991, titled "What would you like to see most in minix?" It began:

> Hello everybody out there using minix—
>
> I'm doing a (free) operating system (just a hobby, won't be big and professional like gnu) for 386(486) AT clones. This has been

brewing since april, and is starting to get ready. I'd like any feedback on things people like/dislike in minix, as my OS resembles it somewhat (same physical layout of the file-system (due to practical reasons) among other things).

I've currently ported bash(1.08) and gcc(1.40), and things seem to work. This implies that I'll get something practical within a few months, and I'd like to know what features most people would want. Any suggestions are welcome, but I won't promise I'll implement them:-)

Linus (torvalds@kruuna.helsinki.fi)

PS. Yes—it's free of any minix code, and it has a multi-threaded fs. It is NOT [portable] (uses 386 task switching etc), and it probably never will support anything other than AT-harddisks, as that's all I have:-(. [52]

At the time he made his efforts public, Torvalds had not yet given the kernel a name, asked other people to help him with development, or even shared his code publicly. (That did not happen until September 17, 1991, as the next section explains.) Yet his plan to build "a better Minix than Minix"—an important development because his attitude had originally been that his kernel "didn't have to do more than Minix"—was clear in his request for suggestions from the Minix community regarding features that were absent in Tanenbaum's operating system, but which they would like to see implemented.[53] That made Torvalds's kernel compelling. It encouraged contributions to Linux from other programmers, whose ranks steadily grew, along with the Linux code base.

Before examining how Linux evolved after other programmers began collaborating with Torvalds, it is worth reflecting on the significance of his achievement in writing an early version of the kernel without outside assistance. Some members of the FOSS community, such as those who have called Torvalds a "god," have perhaps overstated how heroic his single-handed kernel-development effort was.[54] As Tanenbaum noted in his essay on Minix history, Torvalds was hardly the first programmer to write a Unix-like kernel with minimal assistance from other people. On the contrary, by Tanenbaum's count, "five people or small teams had independently implemented the UNIX kernel or something approximating it" before Torvalds began his work.[55] Moreover, as complex as a monolithic kernel is, the amount of coding involved in writing one is relatively small compared to the work required to produce all of the other programs that comprise a full operating system. Even by the late 1990s, when Linux had grown into a mature, feature-rich kernel that contained many times as much code as the primitive version Torvalds completed on his own in 1991, Linux accounted for only about 3 percent of the total code in a GNU/Linux operating system. GNU utilities and programs constituted about ten times as much.[56]

Yet Torvalds's ability to transform his Minix terminal emulator into a complete kernel in a matter of months was no mean feat. Torvalds was not the first programmer to write a kernel single-handedly, but his predecessors in that role had all been professional developers or computer science professors. In contrast, Torvalds had barely a year of college education behind him when Linux debuted. In addition, Torvalds's predecessors had access to many resources that he did not, including powerful hardware, sophisticated debugging tools, and on-site colleagues with whom they could share ideas. In contrast, Torvalds worked from a low-cost 386 computer in his Helsinki apartment, using only the coding tools available to him on Minix—which included many vital GNU utilities but not the extensive software resources of an institution such as Bell Labs, MIT, or Berkeley. In this respect, Torvalds's ability to produce a working kernel was truly remarkable—so much so that, as the next chapter explains, Microsoft-funded researchers in the early 2000s accused Torvalds of plagiarizing the Linux code by claiming that it was simply impossible for him to have produced all of it himself.

## LINUX GROWS UP

On September 17, 1991, about three weeks after Torvalds announced his kernel project on Usenet, the first version of the software, 0.01, became publicly available. The release arrived thanks to Ari Lemke, an administrator at the Helsinki University of Technology, who posted the Linux source code to the university's FTP server, where anyone could download it freely.[57]

In a decision that exerted a much more enduring effect than anything in the actual Linux 0.01 code, Lemke uploaded the source into a directory named *put*/OS/*Linux* on the FTP server he administered. That proved to be important because users came to know the kernel by the name *Linux*, which was not the one Torvalds initially intended to give his software publicly. Although Torvalds had stored the kernel code on his own computer in a directory called *linux*, he planned to release it to the world under the more jocular and modest name *Freax*—a combination of the words *free* and *freaks* that ended in the letter *X*, like the names of most other Unix-like operating systems. "I didn't want to ever release it under the name Linux because it was too egotistical," Torvalds wrote after the fact. Yet *Linux* was the name that stuck.[58]

Torvalds did not publicly announce the availability of Linux 0.01 when it appeared on Lemke's server in September. "Instead, I just informed a handful of people by private email, probably between five and ten people in all," he wrote in his autobiography.[59] Meanwhile, he worked on transforming Linux into a standalone kernel that was capable of running a complete development environment. He originally used Minix as the platform on which to write the Linux code, but after destroying his Minix installation by accidentally writing data to the device /dev/hda1 (his computer's hard disk) instead of /dev/tty1 (a terminal), he decided to discard Minix and never look back.[60]

The work toward a standalone kernel was not quite complete when, on October 5, 1991, Torvalds announced the release of Linux 0.02 in his next major Usenet post, this one written with strong conviction and purpose—not to mention Torvalds's characteristically impressive command of the English language. It began as follows:

> Do you pine for the nice days of minix-1.1, when men were men and wrote their own device drivers? Are you without a nice project and just dying to cut your teeth on a OS you can try to modify for your needs? Are you finding it frustrating when everything works on minix? No more all-nighters to get a nifty program working? Then this post might be just for you:-)[61]

The post went on to describe the technical features and limitations of Torvalds's kernel, including the requirement that users have a Minix system because Linux was still not yet a "standalone" kernel capable of running on its own. Linux also worked only with certain types of hard disk, which not all potential users possessed. Lastly, Torvalds warned, "You also need to be something of a hacker to set it up," emphasizing that Linux was not at all ready for the masses. If users sought a stable Unix-like operating system for 386 computers, he wrote, they should stick with Evans's Minix port.

Torvalds also thought it necessary, when announcing Linux 0.02, to justify his project. Previously, when Linux remained only a personal endeavor, he had not bothered explaining to other people why he thought the kernel was worth developing. In the October 1991 post, however, he declared:

> I can (well, almost) hear you asking yourselves "why?" Hurd will be out in a year (or two, or next month, who knows), and I've already got minix. [Linux] is a program for hackers by a hacker. I've enjouyed [sic] doing it, and somebody might enjoy looking at it and even modifying it for their own needs. It is still small enough to understand, use and modify, and I'm looking forward to any comments you might have.

In Torvalds's own words, the fact that the Hurd was not yet ready to use—and as he joked in another Usenet post a couple of months later, might not appear for "the next century or so:)"—was one justification for writing Linux that would make it appealing to other programmers.[62]

However, the most important part of Torvalds's pitch was that his kernel was written "for hackers by a hacker." He implied in the post that this characteristic was the chief distinction between Linux and Minix. He did not elaborate on exactly what made a hacker's kernel different from the alternatives. But his thinking on this point came through clearly enough in other parts of the Usenet post where he emphasized that Linux was free to use and its source code was fully redistributable. The post did not mention what Torvalds believed to be the technical deficiencies in Minix as a reason for other programmers to consider using Linux. The message was instead about offering the hacker community something it lacked.

In addition to explaining why Linux might appeal to other hackers, Torvalds used the 0.02 release announcement to solicit their feedback and help for developing the kernel further. "I'm also interested in hearing from anybody who has written any of the utilities/library functions for minix," he wrote. "If your efforts are freely distributable (under copyright or even public domain), I'd like to hear from you, so I can add them to the system."[63]

On this occasion (unlike when Torvalds had requested information about POSIX standards the previous summer), helpful responses proved forthcoming. At first, hackers wrote to Torvalds to offer only "maybe one-line bug fixes."[64] By November 1991, however, many people were emailing him to help develop Linux. These included programmers who sent code to implement new features in the kernel and others who merely troubleshot what Torvalds had written himself. "Each day, the community of Linux users expanded, and I was receiving email from places that I'd dreamed about visiting, like Australia and the United States," Torvalds wrote of the period in late 1991.[65]

Collaborators and supporters whom Torvalds never met in person were also offering cash gifts. Such donations were in step with the tradition of the PC shareware culture of the time, in which it was common practice to send the author of a software program who gave away his work for free something on the order of $10 or its equivalent as a thank-you.[66]

Torvalds, however, responded to the offers of cash by asking for postcards instead.[67] "I didn't want the money for a variety of reasons," he wrote in his autobiography. One was that receiving praise and feeling that he was collaborating with a community of researchers—even though he had never held any sort of title or formal status as a researcher himself—was more important to him than material reward. "When I originally posted Linux, I felt I was following in the footsteps of centuries of scientists and other academics who built their work on the foundations of others—on the shoulders of giants, in the words of Sir Isaac Newton," Torvalds wrote in 2001. "Not only was I sharing my work so that others could find it useful, I also wanted feedback (okay, and praise)."[68]

Torvalds also explained his aversion to cash donations by noting, "I suppose I would have approached it all differently if I hadn't been raised in Finland, where anyone exhibiting the slightest sign of greediness is viewed with suspicion, if not envy."[69]

Whatever the root of his desire to avoid financial compensation for his work on Linux, the decision underlined how committed Torvalds was to keeping Linux a completely free operating system, distinguishing it from Minix and most other Unix-like platforms of the time. Even voluntary cash donations, in his mind, would have sullied his project.

The enthusiastic responses that Torvalds received for his work on Linux were not universal. The platform had its critics, chief among them Tanenbaum, the computer science professor whose work on Minix had been an important stepping stone for Torvalds in writing his kernel. On January 29, 1992, Tanenbaum publicly berated Linux as an "obsolete" kernel, primarily because of its monolithic design. Chiding Torvalds for being a lowly student rather than a credentialed computer science researcher like himself, Tanenbaum wrote, "among the people who actually design operating systems, the debate is essentially over. Microkernels have won."[70]

Tanenbaum also criticized Torvalds for having written Linux specifically to run on Intel's 386 computer chips. Tanenbaum wrongly believed these processors would not remain an important part of the hardware market. To Torvalds, who had written Linux on a 386 computer because that was all he happened to have, the attack probably seemed cruel, coming as it did from a computer science professor who faced no such constraints obtaining access to different, more expensive hardware platforms.

"Don't get me wrong, I am not unhappy with LINUX," Tanenbaum concluded. "It will get all the people who want to turn MINIX in[to] BSD UNIX off my back. But in all honesty, I would suggest that people who want a **MODERN** 'free' OS look around for a microkernel-based, portable OS, like maybe GNU or something like that."[71]

For Torvalds, Tanenbaum's public attack touched a sensitive nerve, particularly because it threatened "my social standing" within the hacker community.[72] He responded to Tanenbaum in a Usenet post that made the affair personal, sarcastically writing that, as "a professor and researcher," Tanenbaum had "one hell of a good excuse for some of the brain-damages of minix." He added, "I can only hope (and assume) that Amoeba," another operating system that Tanenbaum had helped to develop, "doesn't suck like minix does." Torvalds enumerated the reasons that he found monolithic kernels to be superior to a microkernel like Minix. He also argued that Linux would prove easier to port to other hardware platforms than Minix because Linux conformed to POSIX standards.

The sparring between Torvalds and Tanenbaum continued for several days. But Torvalds apologized to Tanenbaum for having initially written with "no thought for good taste and netiquette." Other users of Minix and Linux weighed in as well, and a majority supported Torvalds. Although some Linux sympathizers agreed with Tanenbaum that microkernels were superior, several stated that they would still opt to use Linux over Minix because the former was free of cost and available with full source code. "There are really no other alternatives other [sic] than Linux for people like me who want a 'free' OS," one user informed Tanenbaum. In the words of another, "for many people Linux is the OS to use because it's here now, is free and works." The writer added that another benefit of Linux was that it was possible to run the kernel "without paying $$ and/or asking permission from someone."[73]

Neither Tanenbaum nor Torvalds won the debate. Their final correspondence occurred near the end of the year, when Tanenbaum emailed Torvalds to notify him that someone had posted an ad in *Byte* magazine advertising a commercial version of Linux. Tanenbaum asked if that was the sort of thing Torvalds wanted to happen to his kernel. "I just sent him an email back saying Yes, and I haven't heard from him since," Torvalds wrote in 2001.[74]

The criticism that Linux faced from some quarters did not stunt its rapid growth. On the contrary, thanks in large part to the help Torvalds received from fellow programmers who contributed code over the Internet, it gained features and functionality rapidly. By the second week of October 1991, even before Torvalds had completed the transition to making Linux a standalone system, it was capable of running a variety of GNU programs, including the GNU compilers, assembler, make, tar, bash, sed, and awk and a version of Emacs.[75] This was an important step because it meant that Linux had become a kernel capable of supporting a basic distribution of GNU's software suite.

By January 1992, when Torvalds released version 0.12 of the kernel, it had gained page-to-disk support, which allowed the operating system to use hard-disk space to supplement a computer's physical RAM memory.[76] This feature, which Torvalds implemented in response to a request from a Linux user in Germany who wanted to be able to execute code that required more physical memory than his computer contained, was the first that truly set Linux apart from Minix because it added low-level functionality that was entirely absent from Tanenbaum's kernel.[77] Linux 0.12 became "the first release that started to have 'non-essential' features, and [that was] being partly written by others," Torvalds recalled in 1992. "It was also the first release that actually did many things better than minix, and by now people started to really get interested."[78]

More interest fueled an even faster pace of development. By the spring of 1992, a mere eight months after Torvalds had announced on Usenet that he was writing a Unix-like kernel, Linux 0.95 debuted, bringing with it support for graphical applications via the X Windows System.[79] The version number suggested that release 1.0 of the kernel, which would signal that

Torvalds deemed it sufficiently mature for production-level use, was tantalizingly close.

But the path to Linux 1.0 proved more difficult than Torvalds envisioned. By the spring of 1992, the only core component of the kernel that developers had yet to implement fully was support for networking, which would make it possible for computers running Linux to connect to the Internet. Although Torvalds did not think that it would take a great deal of time to develop the networking code, it was not complete until late 1993.[80]

Early the next year, with the networking support in place, Linux 1.0 finally debuted. The University of Helsinki, where Torvalds was still studying and working, organized a launch event. "We got access to the auditorium, and the head of the CS department gave a speech, and all this gave us enough credibility that there was a fair bit of interest from mainstream media," recalled Lars Wirzenius, one of Torvalds's friends and a fellow programmer: "There was even a television crew, and the footage is occasionally found in various places on the Internet. During the speeches, we had a ceremonial compilation of the 1.0 kernel running in the background."[81]

As the Linux code base expanded, so did the kernel's user base. Complete quantitative data on the number of people running Linux during this time period is elusive (as it remains for most segments of the FOSS world today) because most FOSS software does not require purchase or registration, which makes tracking users difficult. Nonetheless, Torvalds told reporters in March 1994, "I guesstimate a user base of about 50,000 active users: that may be way off-base, but it doesn't sound too unlikely."[82]

## COMMERCIALIZING LINUX

The expansion of commercial activity related to Linux even before the release of version 1.0 was another sign of the kernel's rapid adoption. "During 1992 the operating system graduated from being mostly a game to something that had become integral to people's lives, their livelihoods, commerce," Torvalds noted.[83] "There were all these budding commercial companies that had started to sell Linux."[84]

With the 1.0 release, active efforts to promote Linux emerged. These developed organically within the growing Linux community and did not involve Torvalds himself, who remained chiefly interested only in development of the kernel.[85] At the same time, by January 1994, at least one company, Connecticut-based Field Technology, Inc., had begun "selling 'Linux machines' using only copylefted & public domain software," the GNU project reported to its followers. It added, "The Unix-compatible systems are shipped ready to run, with popular programs such as TeX, Emacs, GNU C/C++, the X Window System, & TCP/IP networking. Field Technology makes a donation to the Free Software Foundation for each system sold."[86]

By 1994, the commercial potential of Linux had become so clear that one opportunistic entrepreneur from Boston, William R. Della Croce Jr., who had played no role in Linux development and was unknown to the people involved in the project, attempted to steal the Linux trademark. Initially, the Linux community had not secured a trademark for the operating system's name because, as Torvalds explained in the spring of 1994, "nobody really found the thing important enough to bother about (especially as it does require both some funds and work)."[87] That created an opening that Della Croce exploited on August 15, 1994, when he quietly filed for Linux trademark rights in the United States.[88] The action did not come to the attention of the Linux community until early 1995. When Torvalds and other developers realized that someone with no connection to Linux development might use a legal trick to take over the project, "there was some panic," Torvalds recalled.[89]

Because the Linux community at the time lacked the organizational structure and cash that would be necessary to respond to Della Croce, a consortium of companies with a stake in Linux's future pooled their resources to fund Linux International, a nonprofit association that Patrick D'Cruze had founded to promote Linux before the trademark affair. Under the direction of John Hall, Linux International took the trademark case to court and, in August 1997, settled with Della Croce. The latter agreed to transfer the Linux trademark to Torvalds himself, who, as "benevolent dictator" of the Linux community, promised not to use his control of the trademark to thwart Linux development or business related to it.[90]

Settling the trademark case in the United States did not protect Linux in other countries, however. As Hall reported, "all around the world people were getting the same strange idea." He added that Linux supporters "can't afford to go to the 200 countries around the world and buy trademarks, and maintain them, so we have to fight them on a case by case basis."[91] That is what happened. By 2007, Linux International had spent $300,000 protecting the Linux trademark from prospectors around the globe.[92] Despite such challenges, Linux managed to thrive in the business world during the 1990s, as the next chapter explains.

## THE ART OF LICENSES

When Torvalds released the first versions of the kernel, Linux's commercial success seemed not only improbable but flatly impossible. That is because he initially protected his code, staring with Linux 0.01, with a copyright notice that stipulated the following:

> This kernel is (C) 1991 Linus Torvalds, but all or part of it may be redistributed provided you do the following:
>
> — Full source must be available (and free), if not with the distribution then at least on asking for it.
> — Copyright notices must be intact. (In fact, if you distribute only parts of it you may have to add copyrights, as there aren't (C)'s in all files.) Small partial excerpts may be copied without bothering with copyrights.
> — You may not distribute this for a fee, not even "handling" costs.
>
> Mail me at "torvalds@kruuna.helsinki.fi" if you have any questions.[93]

The original Linux license "was just me writing things up," Torvalds explained in 2016, adding that "there was pretty obviously no actual lawyerese or anything there."[94] The crude copyright amounted mostly to a less sophisticated version of the GPL. It required users of the kernel to accompany distributions of the software with full source code and to adhere to the same licensing terms if they modified Linux. Yet it differed in a key respect. Torvalds forbade distributors of Linux from profiting in any way from the kernel—even if the charges they imposed were intended simply to recoup the "handling costs" they incurred.

These terms reflected how profoundly opposed Torvalds was in 1991 to the idea of charging money for software.

By early 1992, however, Torvalds was rethinking the Linux copyright. Spurred on by Linux users who distributed the kernel at trade shows on floppy disks and asked Torvalds if they could charge nominal fees to cover the cost of their materials, he came to believe that "as long as people gave access to source back, I could always make [Linux] available on the internet for free, so the money angle really had been misplaced in the copyright" that he originally used.[95] He grew even more comfortable with the idea of allowing the sale of Linux as the community and code base grew, instilling confidence that "momentum had been established" and developers "couldn't possibly veer away from our trajectory," Torvalds recalled in his autobiography.[96]

As Torvalds considered adopting a new license for Linux, the GPL was the obvious option. There were other popular licenses in the free software community at the time, such as the permissive ones that governed BSD. But the GPL most closely resembled the homegrown license Torvalds had originally used, and the GPL governed most of the tools Torvalds had relied on to produce Linux. Beginning with the release of Linux 0.12 in January 1992, Torvalds adopted the GPL for the kernel.[97]

Still, despite the GPL's similarity in intent to the original Linux license, Torvalds has never expressed deep satisfaction with it. That is not because he deems the GPL and GNU too constraining. On the contrary, as he noted in 1992, Linux's original copyright notice "was in fact much more restrictive than the GNU copyleft" because of the clause preventing people from making money off of the code in any way.[98] Nonetheless, he has criticized the "hard-core GPL freaks, who argue that every new software innovation should be opened up to the universe under the general public license."[99]

Torvalds also continued for years after adopting the GPL to express concern that someone might charge for Linux or be punished for obtaining a commercial distribution without paying. As he wrote in 2001:

> Generally speaking, I view copyrights from two perspectives. Say you have a person who earns $50 a month. Should you expect him or her to pay $250 for software? I don't think it's immoral for that person to illegally copy the software and spend that five months' worth of salary on food. That kind of copyright infringement is morally okay. And it's immoral—not to mention stupid—to go after such a "violator." When it comes to Linux, who cares if an individual doesn't really follow the GPL if they're using the program for their own purposes? It's when somebody goes in for the quick money—that's what I find immoral, whether it happens in the United States or Africa. And even then it's a matter of degree. Greed is never good.[100]

Even though Torvalds came to believe in 1992 that it made pragmatic sense to adopt the GPL in order to make it possible for distributors to charge money for Linux under certain circumstances, his abiding unease with software that costs money has never fully dissipated.

Ultimately, however, although Torvalds admitted in 1994 that "I'm not fanatic[al] about the GPL," he recognized that "in the case of linux it has certainly worked out well enough."[101] He attributed that success in large part to the fact that, in the Linux world, cooperating with the copyright terms and the development community provides benefits that outweigh those of attempting to subvert the GPL's terms. According to Torvalds,

"It is the people who actually honor the copyright, who feed back their changes to the kernel and have it improved, who are going to have a leg up. They'll be part of the process of upgrading the kernel. By contrast, people who don't honor the GPL will not be able to take advantage of the upgrades, and their customers will leave them."[102]

## GNU AND LINUX

Torvalds's decision to license his kernel under the GPL aligned the code in an important way with the GNU project. In other respects, however, harmony between the Linux community and GNU developers remained elusive during the 1990s.

Part of the unease between the GNU and Linux camps stemmed from the rapid rise of commercial activity related to Linux by fledgling companies whose operations, in the eyes of some hackers, threatened to tarnish the image of free software as the refined, sophisticated endeavor that GNU was pursuing. Ian Murdock, a strong supporter of both GNU software and the Linux kernel, noted that in the early 1990s, "You'd flip through Unix magazines and find all these business card-sized ads proclaiming 'Linux.' Most of the companies were fly-by-night operations," which produced embarrassing implementations of Linux-based systems.[103] From this perspective, Linux appeared to be a poor tool for helping to fulfill the GNU vision of freeing Unix and saving hacker culture.

Such concerns inspired Murdock to float the idea on comp. os.linux (a Usenet group that developers had recently created so that discussions of Linux would have their own home rather than having to use the Minix newsgroup) of building a hacker-friendly operating system that combined the Linux kernel with GNU programs. Stallman responded to Murdock's post and indicated, according to Murdock, that "the Free Software Foundation was starting to look closely at Linux and that the FSF was interested in possibly doing a Linux system, too."[104] With the Free Software Foundation's support, Murdock in 1993 began building Debian GNU/Linux, which today remains a popular Linux and GNU-based operating system.

The partnership between the Free Software Foundation and Murdock surrounding Debian was significant because, until that time, Stallman and other GNU developers had shown little interest in Linux. That was due partly to the fact that, until Torvalds adopted the GPL, the Linux kernel was no more useful to the GNU project than BSD's alternative or any other software that could not be distributed under terms that the Free Software Foundation approved. Stallman also placed little stock in Linux early on because a friend who had reviewed Torvalds's code concluded that it would be difficult to port the kernel to hardware other than the Intel 386 platform.[105] Because GNU developers in the early 1990s remained uninterested in supporting microcomputers, as the previous chapter notes, a kernel that ran only on 386 PCs was of little relevance in their eyes. For them, Hurd remained the only obvious solution to implementing a complete free software operating system.

Gradually, however, the Free Software Foundation's interest in Linux grew. The GNU newsletter mentioned Torvalds's kernel for the first time in June 1992 in a brief paragraph within a long section deep in the newsletter containing updates on free software for microcomputers. Describing Linux as a "free Unix system for 386 machines" that was "named after its author, Linus

Torvalds," the newsletter went on to note the kernel's ostensible limitations—that it "runs only on 386/486 AT-bus machines, and porting to non-Intel architectures is likely to be difficult as the kernel makes extensive use of 386 memory management and task primitives." GNU developers did, however, inform readers which servers they could connect to in Europe and the United States to download the Linux code. The notice did not explicitly mention that Linux was licensed under the GPL, but it implied as much by calling it "free."[106]

The GNU newsletter continued to provide basic information on Linux for the next two years, but not until June 1994 did Linux begin featuring more prominently in GNU's announcements. At that time, the project informed followers that GNU developer Arnold Robbins would be authoring a regular column called "What's GNU?" in the new *Linux Journal*, thereby helping to ensure that GNU's name would remain a part of the discussion surrounding Linux.[107]

The launch of the Debian GNU/Linux distribution, which GNU developers pitched in 1994 as "a complete, full-featured system based on GNU and Linux that is easy to install and configure," also strengthened the Free Software Foundation's role in the evolving Linux world.[108] The relationship between Debian and the Free Software Foundation expanded in the spring of 1995, when the Debian distribution on CD-ROM became available through GNU's official software distribution service.

The Free Software Foundation also took clear note by 1995 of the efforts that were underway by that time to port Linux to hardware platforms other than the 386. The organization announced to GNU newsletter readers in January that a port of

Linux to the Motorola m68k architecture for Amiga and Atari computers was already in testing, adding that versions for the AlphaPC and MIPS processors were in progress.[109] GNU developers' awareness of Linux's growing suitability for a broader range of hardware platforms suggests that, by this time, they were giving the kernel more serious consideration as a long-term stand-in for the Hurd.

By early 1996, their position on Linux had grown even firmer. The January GNU newsletter noted, "The Hurd is not yet ready for use, but in the meantime you can use a GNU/ Linux system."[110] Hurd remained the centerpiece of the GNU vision, but Linux had emerged as an increasingly appealing substitute. Since, as the previous chapter notes, Hurd development was never completed, Linux remains the kernel that powers most GNU-based operating systems today.

Despite early friendly cooperation, relations between the Free Software Foundation and the Linux community grew more tense starting in the summer of 1996. The shift resulted from decisions by both sides. For one, the Debian project changed leadership, with Murdock stepping down and Bruce Perens signing on in his place.[111] Under Perens, the Debian team decided to forgo the sponsorship from the Free Software Foundation that the project had received previously. The reason, Perens said, was that "I decided we did not want Richard [Stallman]'s style of micro-management."[112] GNU developers insisted at the time that they and the Debian team had parted ways "amicably," yet they also stated that they "wish the situation were otherwise" and were considering ceasing to distribute Debian CD-ROMs.[113]

Also in the summer of 1996, Stallman first went on the offensive against usage of the word *Linux* to describe an entire operating system that combined Torvalds's kernel with the GNU suite of programs. He advocated for *GNU/Linux* instead. Murdock had introduced this term, at Stallman's request, to the free software community following Debian's launch.[114] (Stallman had initially suggested the word *Lignux*, but hackers rejected that idea.)[115] Whether out of ignorance, haste, or active hostility toward the GNU project, however, some developers and users of free software referred to GNU/Linux distributions simply as *Linux*. That was a problem, Stallman wrote in an essay on the subject in July 1996, partly because such usage failed "to give the GNU Project credit for making the free Unix-like system that it set out for a decade ago." But he added that

> there is a more important reason for friends of GNU to use names like "Linux-based GNU system" instead of "Linux system." This is to help spread the GNU Project's philosophical idea: that there is ethical importance in freeing users to share software and cooperate in improving it; that free software belongs to a community, and people who benefit from the community should feel a moral obligation to help build the community when they have a chance.[116]

Stallman's words made clear that, to him, the greatest danger posed by people who did not afford full credit to GNU's role in making Linux-based operating systems possible was that they undercut the Free Software Foundation's effort to protect the community of hackers for whom Stallman had launched his crusade.

The essay quoted above recalled the 1983 announcement of the GNU project, when Stallman said that the "Golden Rule" required developers to share source code. But the language about ethics and moral obligations that Stallman deployed in 1996 elevated his rhetoric to a new level. For many supporters of free software, Stallman was laying the ideology on too thickly. Ultimately, stances such as the one he took in the 1996 essay contributed to the fracturing of the hacker community into "free software" and "open source" camps, which chapter 5 details.

The Free Software Foundation continued over the following years to denounce what it saw as betrayals of the true purpose of free software. In 1997, Stallman lamented in an essay that some supporters of Linux did not endorse the free software philosophy or the GPL at all:

> A conference was held this year on the topic of developing "Linux applications." This conference was about using the GNU system, but the conference announcement did not mention the word GNU. Instead of encouraging users to write more free software, it did just the opposite. It included a panel entitled, "Licenses and licensing—I don't want to give away my application!!!"[117]

By the spring of 1998, Stallman had come to accept that Linux was likely to remain the kernel that most people would use to build free operating systems for the foreseeable future. He acknowledged that year that completing the Hurd had proved "a lot harder than we expected, and we are still working on finishing it," but added, "fortunately, you don't have to wait for it, because Linux is working now. When Linus Torvalds wrote Linux, he filled the last major gap. People could then put Linux together with the GNU system to make a complete free system: a Linux-based GNU system (or GNU/Linux system, for short)."[118]

In the same essay, Stallman remained harshly critical of people who left the *GNU* out of *GNU/Linux*. Indeed, he went on the counterattack by suggesting that the word *GNU* alone might fairly be used to describe operating systems that included the Linux kernel. That usage, he argued, was valid because statistics showed that GNU software accounted for nearly ten times as much of the total source code of a GNU/Linux distribution of the time as the Linux kernel. "So if you were going to pick a name for the system based on who wrote the programs in the system," he concluded, "the most appropriate single choice would be 'GNU.'"

Significantly, in this essay—which Stallman wrote just as the term *open source* entered hackers' lexicon and some major free software developers began distancing themselves publicly from the Free Software Foundation—Stallman no longer portrayed correct usage of *GNU/Linux* as a moral or ethical issue. His main argument for giving full credit to GNU developers for their software was instead that they had done most of the work required to make possible a free operating system that used the Linux kernel. Following public suggestions that the Free Software Foundation had become overly ideological, Stallman seemed to decide to temper his rhetoric for strategic purposes.

If that was the case, the toning down of Stallman's rhetoric occurred too late. As chapter 5 shows, the split between the free software community that Stallman led and the open source group that grew up around Torvalds was too divisive by the spring of 1998 to have a clear resolution. Even worse from the perspective of the GNU camp, observers of the Linux revolution would soon be calling Stallman a "forgotten man."[119]

## WHY DID LiNUX SUCCEED?

Why did Stallman and the GNU developers, who had breathed the first life into the FOSS movement and written most of the code that made FOSS possible, end up on the sidelines as Linux exploded in popularity? How did a cheeky undergraduate end up—in the eyes of many hackers, at least—inheriting the fruits of the revolution that Stallman had launched?

These were outcomes that few people could have predicted. In August 1991, it would have been almost absurd to expect that the kernel Torvalds had just announced to the world would ever find a large following outside his Helsinki apartment or power millions of computers across the planet. Other people—including Tanenbaum, the BSD programmers, and the GNU developers—were in the process of building freely redistributable Unix-like kernels or had already completed them. By the standards of the time, their kernels were much more technically impressive than the one Torvalds aimed to create. Those other programmers also had many more credentials to their names than the Finnish undergraduate who created Linux, and they enjoyed immensely larger development budgets.

It was telling that Salus's monumental study of the history of Unix-like operating systems, which he published in 1994, devoted only a handful of sentences to the Linux kernel.[120] As late as that year, even someone as tuned into the ecosystem of Unix-like operating systems as Salus did not foresee how important Linux would become. Yet barely five years after Torvalds announced his project on Usenet, Linux had evolved into the kernel that, by all realistic measures, had beat out all of the better-funded, more elaborate alternatives.

Historically, most observers of the FOSS community have attributed Linux's success to fortunate timing. Wirzenius, the Finnish programmer who shared an office with Torvalds at the University of Helsinki, wrote in 1998, "The success of Linux wasn't automatic, and things might well have gone differently. For example, if the Hurd had been finished a few years ago, Linux probably wouldn't exist today. Or the BSD systems might have taken over the free operating system marketplace."[121] Tanenbum made a similar point when he noted that the legal troubles surrounding BSD "gave Linux the breathing space it needed to catch on."[122]

To a significant extent, such interpretations account for Linux's improbably rapid and widespread growth during the early 1990s. Uncertainty over the legality of kernels derived from the BSD code base stunted their adoption, creating an opening for Linux that might not otherwise have been so broad. Meanwhile, as Torvalds himself noted, he likely would not have bothered to develop Linux at all if the Hurd had been closer to completion or if Minix had worked better on his PC.

Yet good timing alone does not fully account for Linux's ability to bloom during its first years. Although BSD's legal battles were significant, the Unix Systems Labs lawsuit did not begin until January 1992. By that time, Linux already had a small following, which extended beyond Torvalds's personal circle. Moreover, the BSD case was settled at the start of 1994, just as Linux 1.0 made its debut. Linux at that time was no more sophisticated in technical terms than the free BSD kernels, which at any rate supported a broader range of hardware platforms. Plus, for most users BSD and its derivatives were also just as free of cost as Linux. The legal troubles between 1992 and 1994 certainly helped to push developers and users away from the BSDs and toward Linux, but the differences between these two options were not great enough to make Linux the only obvious choice by the time the dust had settled in the courtrooms.

On a similar note, Linux faced its own legal uncertainties during the trademark case that began in 1994 and lasted until 1997. Della Croce's attempt to usurp the Linux trademark was not as dangerous for the Linux kernel as the challenges over code ownership were for BSD. But it was still a serious concern. Despite this fact, it did not lead to a massive desertion from the Linux community in the years before Torvalds obtained the Linux trademark. It did not even stop companies from taking a serious commercial interest in Linux. This outcome showed that there was something intrinsically different about Linux, which emerged relatively unscathed from the same genre of legal trouble that had placed the BSD community in crisis.

What was that difference? Above all, licensing. The BSD license was in a sense even freer than Linux's GPL. As noted above, the BSD licensing terms allowed developers essentially to do whatever they wished with code derived from BSD's code base, even if they did not make the source code of derivative software publicly available. For that reason, BSD might have seemed more attractive for developers than Linux because it was more flexible. Yet because the FOSS camp's chief goal was to nurture hacker values, not develop software with as little fuss as possible, the GPL-licensed Linux kernel, which respected hacker mores, proved more attractive than BSD code to many hackers.

It mattered, too, that the Linux kernel code never cost a penny. That set it apart from most of the other freely licensed Unix-like kernels of the early 1990s. NET 2, the first complete BSD-based operating system that did not require users to purchase a Unix license from AT&T, could be legally copied and distributed by individuals without paying, but an official copy from Berkeley itself cost $1,000.[123] Some of the other BSD derivatives were similarly priced.[124] Minix cost a fair amount of money, too, as Torvalds bitterly noted.

The role of cost in Linux's success should not be overstated. By 1992, several BSD-based systems that cost no money, no matter how users obtained them, were in circulation. And again, it was always possible for users to obtain NET 2 without cost if they copied it themselves. Yet Torvalds's adamant opposition during Linux's early days to charging money for the kernel made his project a true outlier. It also probably augmented Linux's appeal for hackers who, having witnessed the troubles that ensued when AT&T commercialized Unix in 1983, were wary of any activities by software distributors that smacked of commercialism—even if the distributors also offered ways to obtain code without paying.

Lastly, the developer community that Torvalds cultivated helped to ensure the kernel's rapid growth. To an extent, as noted above, programmers were attracted to Linux because of uncertainty over BSD's legal future, which sapped the BSD community of its momentum and made some programmers wary of contributing code. Yet programmers who liked Linux's nonexistent price tag and the openness of its code began helping Torvalds write the kernel before the BSD lawsuits commenced. On October 10, 1991, several months before the start of the first BSD case, Torvalds wrote that his kernel "never would have seen the light of day or would have been much worse without the help of some others."[125] He went on to name collaborators who were helping him develop Linux via the Internet. By the time of the Linux 0.11 release in late 1991, "a small following" of programmers had arisen, according to Torvalds.[126] From there, the community of Linux developers continued to grow.

Torvalds was not the first programmer to endorse a decentralized, Internet-based community of developers. Keith Bostic, a lead BSD developer, did something similar in 1990 when he enticed hundreds of volunteer programmers from across the Internet to help rewrite Unix utilities without AT&T code in preparation for the release of NET 2.[127] And according to Tanenbaum, Linux followed "essentially the same development model as MINIX," which also received contributions of code from users who wished to add new features.[128]

Yet Torvalds's project was different in several key ways from other collaborative, Internet-based development efforts. First, because Linux was completely free of cost, programmers who donated code to Torvalds could reap what they sowed without having to pay a penny. They could not do the same in the case of Minix. Second, Linux came of age as email access became widespread and the Internet ceased "being an enclave of a few research universities," as Torvalds recalled.[129] That lowered the barrier to participation.

Third and most important, at least in Torvalds's own view, was that the community of programmers who contributed to Linux early in its history was not based at a single institution or derived from a core group of original participants. "There was no historical insider group," Torvalds told me, "so we were a lot easier to approach if you came from a DOS/Windows background, for example." He added, "there was no cabal, it was easy to send me patches, I wouldn't have stupid paperwork rules like a lot of other projects had, and it really was a much more open project than a lot of software projects that preceded it."[130]

Not until the late 1990s, when Eric S. Raymond published work on the "bazaar" mode of software development, did FOSS developers fully recognize and articulate the novelty of the decentralized, Internet-based Linux development model. And it was only in 2008, with the introduction of GitHub, which reduced barriers to FOSS contribution even more than email-based development had, that open collaboration on FOSS code saw its complete incarnation. Yet the effects of the model Torvalds helped to pioneer were apparent much earlier. They were why observers noted in the spring of 1994, "The number and frequency of new releases of Linux, and drivers and utilities, are amazing to anyone familiar with traditional UNIX development cycles."[131] The approach Torvalds adopted in 1991 remains highly influential today, when most major FOSS projects—from the OpenStack cloud computing platform to the Linux Foundation's various "collaborative projects"—follow the same model.

A variety of factors combined to make Linux into the sophisticated, feature-rich kernel that it became by the mid-1990s. From there, they continued to fuel the remarkable expansion of the Linux ecosystem, which, as the next chapter shows, assumed outsize importance within the computing industry by the end of the decade.

# 4 THE MODERATE FOSS REVOLUTION

ONE OF THE MAIN ACTS in the revolutionary script is the moderate phase. In this part of a revolution, consensus prevails, and competing parties find enough common ground to establish a new, stable order. The moderate phase tends to precede and be less exciting than the dramatic, battle-ridden periods that usually follow. But it is generally during the moderate stage that the most productive and enduring revolutionary changes arise.

For the French revolutionaries, the moderate phase began after Louis XVI bowed to popular demands by accepting constitutional monarchy in the summer of 1789. Over the next few years, the French abolished feudalism, reconfigured church-state relations, and pioneered new modes of representative democracy. After foreign and civil wars began and the king was overthrown and imprisoned in 1792, however, the French Revolution descended into the radical phase known as the Reign of Terror.

The Russian Revolution followed a similar trajectory. A relatively peaceful revolution in March 1917 resulted in a new, provisional government that was dominated by centrist factions. These groups shared power with the more radical Bolsheviks until the latter launched a second revolution in November. The struggle between Red Bolsheviks and their White Russian enemies then enveloped the country in a prolonged civil war, complicated by foreign intervention, which erased hopes for moderate reform.

The free and open source software (FOSS) revolution also had a moderate phase. The early and mid-1990s constituted an eventful yet peaceable period during which different factions within the FOSS community collaborated readily with one another. The unchecked momentum of the Linux kernel and the various programs that GNU developers produced gave rise to operating systems through which FOSS reached the masses for the first time. Meanwhile, new values and methods, like those that grew up around the Apache Group (which in 1999 became the Apache Software Foundation), were integrated into the rapidly expanding FOSS ecosystem. FOSS was sufficiently successful to show that a new world was possible but not yet threatening enough to enter the crosshairs of companies like Microsoft, and it flourished.

This chapter chronicles the moderate phase of the FOSS revolution. It focuses on the advent of GNU/Linux distributions, the proliferation of FOSS productivity applications, and the surging commercial significance that these developments introduced to the FOSS world. It also highlights the beginnings of the fissures that started emerging in this period. Despite a general spirit of consensus, opposing factions began staking competing claims to the FOSS revolutionary legacy by adopting different types of licenses and development strategies. FOSS programmers also began attacking closed source software companies loudly for the first time.

# FREE NEW WORLD: GNU/LINUX DISTRIBUTIONS IN THE 1990S AND EARLY 2000S

## The First GNU/Linux Distributions

The code that Ari Lemke posted on a University of Helsinki file transfer protocol (FTP) server in September 1991 represented the first "distribution" of the Linux kernel. But that was not a complete Linux-based operating system. At the time, Linux included little beyond basic kernel code, and users still required Minix to run the system.

Within less than a year, however, free software supporters began combining the Linux kernel with suites of other utilities, most of them from the GNU project, to build what hackers eventually called GNU/Linux distributions. The first example of such a system came from England, where Owen Le Blanc of the Manchester Computing Centre created MCC Interim Linux, named after the institution where he worked. Le Blanc's distribution was basic. It provided Linux kernel version 0.12 in conjunction with a small set of programming tools and an installer. Nonetheless, it demonstrated that the Linux kernel, though still novel at the time and hardly as promising in appearance as the more sophisticated kernels under development by the GNU and BSD teams, could constitute the basis for a viable operating system.[1]

A more feature-rich distribution appeared several months later in the form of Softlanding Linux System (SLS), which was the first GNU/Linux system to gain a large following. Launched in May 1992, SLS extended free software functionality by integrating the X Windows System and TCP/IP Internet connectivity alongside the Linux kernel and GNU utilities.[2]

The end of 1992 saw the debut of the first commercial GNU/Linux distribution, Yggdrasil. Sold by Berkeley-based Yggdrasil Computing, Inc., and named after a holy ash tree from Norse mythology, Yggdrasil became available for testing in December 1992. The distribution was based in its initial incarnation on Linux kernel version 0.98 and cost $50 in beta form. For the production release, the price increased to $99.[3]

Significantly, Yggdrasil was the first major GNU/Linux distribution to ship on a "live CD." That made it possible for users to run a complete Yggdrasil system in their computer's random access memory (RAM) using only removable media. They also could use the live CD to install Yggrdasil permanently on their hard drives, but making permanent changes to the computer was not a requirement. The Yggdrasil live CD made it easier to test a FOSS system. It was also an attractive feature in an era when other GNU/Linux distributions required as many as seventy-three floppy disks for installation.[4]

### GNU/Linux Grows Up

The first public GNU/Linux distributions were plagued by software bugs and lacked user-friendliness. Demand for software that performed better fueled rapid improvements. In July 1993, version 1.0 of Slackware GNU/Linux appeared. It originated as an effort to create an enhanced, less buggy version of the SLS distribution. Featuring a user-friendly installer named *dialog*, Slackware quickly gained a large following. It remains under active development today, a fact that distinguishes it as the oldest continuously developed GNU/Linux distribution.[5]

Slackware spawned several other important GNU/Linux distributions. These included SuSE, whose name is an acronym for the German term *Software- und System-Entwicklung* (Software and System Development). The SuSE operating system, which first shipped in 1994, was the product of a group that Roland Dyroff, Burchard Steinbild, Hubert Mantel, and Thomas Fehr founded in 1992 to do consulting work, not distribute free software.[6] Although the early SuSE releases were derivatives of Slackware, the SuSE developers in 1996 integrated another GNU/Linux distribution, Jurix, into their platform, helping to form the distinctive SuSE ecosystem that has sustained both a commercial GNU/Linux distribution, SUSE Linux Enterprise, and a community-supported variant, openSUSE, into the present. (Modern versions of SUSE capitalize all letters in the operating system's name, although *SuSE* was used historically.)

Another GNU/Linux distribution that was destined for long-term commercial success, Red Hat, had its humble beginnings around the same time as SuSE. Named after a Cornell University lacrosse hat that the distribution's founder, Marc Ewing, wore as a student at Carnegie Mellon, the Red Hat distribution (though not the company now associated with it) was born in October 1994. Its rise to commercial importance in the late 1990s, which this chapter details below, followed the distribution's acquisition by a company called ACC Corporation, which originally sold software and documentation for Unix and Unix-like systems.[7]

While some entrepreneurs were pursuing commercial opportunities related to Linux-based operating systems, wariness toward such activities prompted other developers to create distributions for the purpose of keeping the software freely available. The most prominent example of this was Debian GNU/Linux. As the previous chapter explains, Debian began as an independent project under the direction of Ian Murdock. The distribution's name was an amalgamation of the first names of Murdock and his girlfriend and later wife, Deb.[8] Billed as "an attempt to create a non-commercial distribution that will be able to effectively compete in the commercial market," Debian was the first operating system based on Linux and the GNU software suite to endorse the name "GNU/Linux."[9] It received support from the Free Software Foundation until November 1995 and was available through GNU's official distribution service.[10]

As the previous chapter notes, the Free Software Foundation and Debian later parted ways due to concerns among Debian's leaders with what they called Stallman's "micromanaging" tendencies. After the split, the Debian team founded a group named Software in the Public Interest to oversee development of the distribution.[11] In its effort to distinguish itself from organizations using Linux for commercial purposes, the Debian project touted its system as being "entirely free to use and redistribute," adding that "there is no consortium membership or payment required to participate in its distribution and development. The developers are 100 unpaid volunteers from all over the world who collaborate via the Internet."[12]

Despite Murdock's death in December 2015, Debian remains actively developed, freely available, and independent of direct ties to commercial parties—although it serves as the basis for other GNU/Linux distributions created by organizations with commercial ambitions.[13] Ubuntu, a younger operating system whose history chapter 6 details, is the most notable example of these.

The distributions discussed above helped to encourage Linux kernel adoption in the early and mid-1990s. By the turn of the new millennium, however, the landscape of GNU/Linux distributions had grown even more diverse. The most interesting newcomers included operating systems designed to integrate the FOSS and proprietary software worlds by allowing users to run Microsoft Windows applications seamlessly alongside GNU programs and other free software, all atop the Linux kernel. In 1999, Corel Corporation, a Canadian company, introduced Corel Linux. Based on Debian, the system included a customized variant of the Wine compatibility layer, the tool for executing Windows binaries on Linux that this chapter discusses below in greater detail. Using Wine, Corel Linux could run Windows programs, including those in the WordPerfect productivity suite, which Corel owned.[14]

Corel Linux intrigued observers in both the FOSS and proprietary software worlds for a time. But ultimately, it failed to entice Windows users to adopt Corel's Windows-friendly GNU/Linux platform or people using a different GNU/Linux distribution to migrate to Corel.[15] By August 2001, Corel Linux folded, although Corel the company remained in business. Xandros, a company that was formed the previous May, purchased the Corel distribution source code and development team and used them as the basis for building its own Linux-based system, which became an important—though not record-breaking—GNU/Linux distribution during the first decade of the new millennium.

Despite Corel Linux's lack of success, similar distributions arose in its wake that also aimed to integrate Windows and FOSS environments. In August 2001, another start-up, Lindows, Inc., was launched in California to develop a GNU/Linux distribution that used Wine to support Windows binaries seamlessly.

To set their system apart from Corel Linux, the developers also produced a one-click application installer, Click-N-Run, which they promoted as the easiest way to add software to a Linux-based platform. The first version of the operating system, called Lindows, appeared in autumn 2001. It enjoyed endorsement by the trade press, although admirers doubted Lindows's ability to compete with Microsoft Windows in a sustainable way.[16]

Microsoft took Lindows GNU/Linux seriously enough to sue Lindows, Inc. in 2002, alleging copyright infringement in the operating system's name. A court dismissed the charges, but Microsoft renewed the legal battle by requesting a second trial. The parties reached a settlement in 2004, when Lindows, Inc., changed its name to Linspire, Inc., in exchange for an undisclosed cash sum from Microsoft.[17]

Despite Lindows's success in the courtroom, the project failed, as did Corel Linux before it, to build a self-sustaining user base. It was not successful in the late 2000s in attempts to forge partnerships with Canonical, the company that develops Ubuntu Linux, and with Microsoft. In July 2008, Xandros acquired Linspire and discontinued development of the operating system the following month.

In addition to Corel Linux and Lindows, the other major commercial GNU/Linux distribution to emerge near the turn of the new millennium was Mandrake, which debuted in July 1998 under the direction of a company named MandrakeSoft. Mandrake's developers, who used Red Hat's distribution of GNU/Linux as the foundation for their operating system, aimed to innovate by enhancing usability.

They enjoyed good success. By 2002, reviewers were hailing Mandrake as "one of the easiest to install and [most] user-friendly Linux distributions … on the market."[18] Ordinary users compared it favorably to other GNU/Linux distributions. "Mandrake is a solid distro and a great tool if you want to actually use your computer to get work done," one Slashdot commenter wrote. "Such a company deserves our support if we ever want to see Linux prosper on the desktop. Debian sure isn't anywhere close to the 'just install it and it works' stage yet."[19]

Yet like many other distributions in the early 2000s, Mandrake was embroiled in legal battles. In Mandrake's case, the challenger was not Microsoft but King Features Syndicate, a media company. King Features alleged that the *Mandrake* name infringed on its trademark of Mandrake the Magician, a cartoon character. The challenge prompted MandrakeSoft to change the name of its GNU/Linux distribution to Mandrakelinux. In 2005, following the acquisition of the Brazilian company Conectiva, the company modified the name again, to Mandriva Linux. A number of distributions appeared under the Mandriva brand from the mid-2000s until the spring of 2015, when Mandriva developers apparently ceased operations.[20]

While distributions such as Mandrake, Lindows, and Corel Linux were launched with the goal of making Linux and GNU-based software easier to deploy and use, another segment of the FOSS camp in the late 1990s and early 2000s moved in the other direction—by building distributions that prioritized not user friendliness but users' freedom to customize software to the maximum extent possible. In this sense, these distributions mirrored Debian GNU/Linux, which, as noted above, had appeared in the earlier 1990s in response to concerns that the free software world was coming under the sway of commercial entities intent on privileging profit-driving usability over software freedom.

Gentoo was the most enduring distribution emphasizing customizability that was born around the turn of the millennium. A derivative of Enoch, this GNU/Linux distribution debuted in December 1999. It stood out from other offerings because rather than shipping precompiled binaries, it provided users with source code for compiling their own Linux-based operating systems. Version 1.0 of Gentoo, named after a species of penguin, appeared in May 2002. Although Gentoo's source-code-only distribution model required users to have a significant amount of technical expertise, the reward for the time and energy they invested was a system that could support a broad range of hardware platforms. In addition, at least in theory, Gentoo boasted excellent performance because compiling from source meant that users could optimize the software for their particular needs.

Gentoo held enough appeal for the FOSS community that, by 2003, observers were calling it "the fastest growing Linux distribution of all time."[21] Ultimately, however, the distribution's momentum failed to endure. Its initial rapid pace of adoption slowed later in the decade. Still, Gentoo remains under active development today and continues to cater to a small community of technically inclined users.

In addition to the major GNU/Linux distributions described above, at least dozens and possibly hundreds of others (depending on how they are counted) appeared during the 1990s and 2000s. In 2001, *Linux User* magazine estimated that about 140 distributions were in existence. In the same year, DistroWatch.com was founded to track free software usage and provide news, and by 2015, it identified eight hundred distinct distributions, some under active development and others long extinct.[22] (Some of the distributions that DistroWatch tracked include operating systems that use kernels derived from BSD rather than Linux, although these represent a small minority compared to the Linux-based systems.) The difference between these two figures probably does not mean that new GNU/Linux distributions have appeared at a more rapid pace since 2001 than they did in the 1990s. Instead, it reflects the lack of the existence of an official body that maintains statistics on the distributions, as well as the subjective nature of defining what constitutes a distinct distribution as opposed to a variant of a distribution that already exists.

Although many GNU/Linux distributions have emerged over the previous twenty-five years, a small—though changing—minority of them have attracted the vast majority of GNU/Linux users. Most distributions, both historically and today, garnered only small followings. More than a few probably have never been used in a serious way by people other than their creators. That is no surprise, given the availability of tools that allow users with only a moderate level of technical expertise to create distributions of their own. One website in 2015 offered a tutorial on "How to Build Your Own Linux Distro" that consisted of only seven easy steps.[23]

### FREE BSDS

A survey of the FOSS landscape in the 1990s and early 2000s would be incomplete without reference to the derivatives of the BSD Net 2 operating system that emerged alongside the early GNU/Linux distributions. Most of the BSD-based systems combined GNU software with the Net 2 kernel and other utilities from the BSD project to implement freely redistributable Unix-like operating systems.

As of 2015, DistroWatch.com identified twenty-six BSD-based distributions that have existed historically or remain under active development. Of these, only three have enjoyed enduring followings over the long term. The first, NetBSD, appeared in the spring of 1993. Its developers emphasized compatibility with a broad variety of hardware platforms (a goal that spawned the project's official motto, "Of course it runs NetBSD") as well as fitness for running Internet-based software.[24] The latter trait helped to make NetBSD popular as FOSS became an increasingly important part of the growing market for Web servers and other Internet platforms.

The second major BSD-based distribution, FreeBSD, debuted in November 1993 as a community-developed operating system. It soon gained commercial support from a company named Walnut Creek CDROM, which distributed the system on disk.[25]

The third long-lasting BSD variant, OpenBSD, originated in 1995 when Theo de Raadt, one of the NetBSD developers, left that project due to disagreements with other programmers and founded his own distribution. The OpenBSD developers distinguished themselves by adopting a purist approach to licensing. They rejected the GPL as too restrictive and preferred, whenever possible, to license components of their operating system under terms derived from the original BSD licenses. This commitment led them to rewrite the GPL-licensed utilities that were present in the operating system in order to make almost all of the software in the distribution compatible with BSD-style licenses.

Even though BSD licensing terms did not require source code to remain publicly available, the OpenBSD team chose to release its code this way by posting it on Internet servers, a practice that was innovative in the mid-1990s. A strong focus on documentation, which was a weak point in many other FOSS projects, was an additional distinguishing feature for OpenBSD, as was extensive attention to software security.[26]

NetBSD, FreeBSD, and OpenBSD all survive into the present, but the various spin-offs that they have spawned over the years have generally not fared as well. Yet even if BSD-based distributions account for only a small fraction of the total FOSS ecosystem, users who have adopted these systems tend to be particularly dedicated to them. Today's BSD-based distributions also attest to the long-lasting influence on the FOSS space of the Berkeley team's efforts to develop a free Unix-like system, even though the main BSD project, as the previous chapter notes, ceased operations more than two decades ago.

## ADVENTURES IN USERLAND: BUILDING FOSS PRODUCTIVITY APPLICATIONS

The proliferation of GNU/Linux and BSD-based operating systems discussed above depended in large part on the steady expansion of FOSS applications and tools. In the early 1990s, the GNU software suite and the Linux and BSD kernels sufficed for building basic operating systems that were suitable for technically inclined users who were comfortable working from a command line. As the decade progressed, however, more sophisticated programs—ranging from enhanced user interfaces to office productivity applications—appeared. They made FOSS a compelling offering to a much wider demographic of users.

### Freeing the Graphical User Interface

The first software to extend the functionality of FOSS systems beyond the bare essentials was X Windows, a framework for building graphical interfaces on Unix-like systems. X Windows—which developers colloquially call *X*—was the successor to W, a display system developed for use on a platform known (confusingly enough) as the V operating system. Seeking a windowing system that would be portable across different Unix-like operating systems and various hardware platforms, programmers at MIT started developing X in 1984. By 1986, the software had spread beyond MIT, which distributed the code under liberal terms that came to be known as the MIT License. The license allowed anyone to reuse the X code, including as part of software packages that were otherwise closed, provided they afforded the same liberties to people to whom they redistributed X. GNU adopted X as the display solution for its system in 1987. By 1988, X had become the standard graphics platform for GNU and most other Unix-like systems. That made it one of the first widely distributed FOSS platforms at a time when much of the GNU software suite remained under development.[27]

Virtually all of the GNU/Linux distributions of the 1990s that provided graphical interfaces adopted the implementation of X known as XFree86. Although some programmers contended that the software architecture behind X (which was designed to deliver graphical displays on both local computers and over a network) was needlessly complex for standalone computer systems and the X developers themselves warned from the project's start that X "is not the ultimate window system," it has remained the graphical backbone of free and open source operating systems into the present.[28]

Only in recent years have FOSS developers begun seriously considering alternatives to X. The most momentous change in this respect occurred in 2010, when Canonical, the company that develops the Ubuntu GNU/Linux distribution, announced plans to replace X with a display server called Wayland. In 2013, Canonical modified its course by announcing that it would create its own display server, Mir, instead of adopting Wayland.[29] Mir remains under development, however, making X the main display server for Ubuntu and most other GNU/Linux distributions as of 2016.

X provides only the backend that developers need to build graphical applications. It does not include the code that draws elements such as windows, toolbars, and animations on the screen. The type of program that does the latter is called a *desktop environment*. By the early 1990s, a variety of desktop environments existed for Unix and Unix-like system. Some, notably the Common Desktop Environment, were commercial products that most supporters of GNU and Linux found uncompelling because they were not freely redistributable. Other graphical applications of the early 1990s were inconsistent in look and feel or incomplete in functionality—which meant that they "quite frankly sucked big time, and hence no non-literate computer user would've touch Linux with a barge pole," as one Slashdot user put it.[30]

These deficiencies prompted FOSS developers beginning in the mid-1990s to build a variety of free desktop environments that offer a consistent, sophisticated graphical experience across an entire GNU/Linux (or GNU/BSD) system. Like the distributions themselves, the free desktop environments that have appeared in the FOSS world over the preceding two decades are too numerous to list exhaustively here. But an overview of the two most important environments—KDE and GNOME—is in order.

KDE, the oldest FOSS desktop environment that remains popular today, was born in 1996 when Matthias Ettrich, a German student, announced on Usenet that he planned to create a new graphical interface for GNU/Linux systems. Describing his frustrations with the mishmash of different menus, windows, and widgets that existed for FOSS systems at the time, Ettrich called for "a modern and common look & feel for all the applications." He explained (in German-inflected English) that the comprehensive nature of the graphical solution he envisioned was "exactly the reason, why this project is different from elder [sic] attempts."[31] Ettrich invited other programmers to join his efforts to create the new graphical platform, which he dubbed the *Kool Desktop Environment*. The project soon came to be known simply as KDE.

In leading KDE development, Ettrich displayed an appreciation for the needs of ordinary end-users that was unusual among the highly technical programmers who populated FOSS-related Usenet groups in the 1990s. That strength helped KDE advance. Yet Ettrich also unwittingly erected a major obstacle to KDE adoption by choosing to develop the platform using Qt, a new programming library. Qt, which was owned by a company named Troll Technology and subject to a proprietary software license, was far more mature than any graphics library that was available at the time under a FOSS-friendly license. And because

Troll Technology allowed Qt's use with applications running on X free of cost, it was possible to license KDE under the GPL and distribute it with free GNU/Linux systems even if the KDE software relied on the Qt library to run. Yet developers like Stallman denounced Qt as a "trap" because using it made FOSS programmers dependent on a proprietary software framework. Ultimately, critics warned, Qt left users at risk of being unable to run free software in the event that Troll Technologies decided to cease supporting the software or made it incompatible with GPL-licensed code—both of which it could do at any time, with no need to consult the FOSS community.[32]

Nonetheless, as Perens of the Debian project noted, "the prospect of a graphical desktop for Linux was so attractive that many users were willing to overlook" the licensing issues associated with Qt.[33] Enough FOSS developers endorsed the KDE project that the desktop environment rapidly matured. On July 12, 1998, the KDE team issued its first stable release, offering users of GNU/Linux systems a graphical experience that greatly surpassed anything previously available from the FOSS community.

Yet even as the KDE team made steady progress building and spreading its platform, the Qt dependency continued to worry some hackers. With GNU's support, two of them, Miguel de Icaza and Federico Mena, started work in 1997 on an alternative desktop environment called GNOME (GNU Network Object Model Environment).[34] (I have been unable to locate evidence that the name was a jab at Troll Technologies, even though trolls and gnomes are not often friends in the fantasy tales that are popular among some programming geeks.) They aimed to create "a free and complete set of user friendly applications and desktop tools, similar to [the Common Desktop Environment] and KDE but based entirely on free software."[35] Instead of Qt, GNOME relied on a GPL-licensed programming library named GTK+, which originally was developed in conjunction with the GNU Image Manipulation Program (GIMP)—a free, cross-platform application for creating and editing images.

GNU developers were so wary of Qt that they did not stop with GNOME, however. Hedging their bets, they also launched the Harmony project, which aimed to create a clone of the Qt library and license it under the GPL. If fully implemented, the Harmony library would have permitted users to run KDE without relying on software from Troll Technology.[36]

The goal of both the GNOME and the Harmony initiatives was to force Troll Technology's hand by obliging it to release Qt under more liberal terms. They succeeded. Threatened by the prospect that the Qt library might become irrelevant to FOSS users either because GNOME outpaced KDE in popularity or because the Harmony library evolved into a complete substitute for Qt, Troll Technology relented. In June 1999, the company released an updated version of the Qt library under terms it developed called the Q Public License, which guaranteed that the source code for derivatives would remain available.[37]

This decision was not satisfying to the Free Software Foundation, however, because the Q Public License and the GPL were not mutually compatible. The organization therefore continued its dual campaigns to provide GPL-licensed alternatives to Qt-based KDE. Its efforts prompted Troll Technology in 2000 finally to release the most up-to-date version of Qt under the GPL.

Some GNOME supporters were underwhelmed by this development, partially because KDE remained dependent at the time on earlier versions of the Qt library that were not redistributable under the GPL. Worse, KDE contained code that had been copied from GPL-licensed software in ways that violated the GPL. For this reason, Stallman warned that work remained to be done before the "Free Software Movement will be able to think of KDE/Qt as a contribution and not as a problem."[38] He promised "forgiveness" to KDE developers who had improperly used GPL-licensed code but only if they adopted the version of the Qt library that was licensed under the GPL.

Stallman's reaction to the new licensing of Qt did little to smooth over the tensions that had arisen within the FOSS community between supporters of GNOME and KDE. Some users agreed that KDE developers should renounce non-GPL code fully to ensure that the goals of the FOSS community were reached. "Proprietary software is like a cancer, even a little will kill over time," one user wrote in support of Stallman's statement about Qt.[39] (Ironically, the writer deployed a metaphor that Microsoft's CEO used the following year to denounce Linux, which is discussed in chapter 5.)

In a sign of growing wariness about views from the Free Software Foundation that seemed overly radical to many people, some FOSS users were critical of Stallman's obstinacy. "Here is a man who get's [sic] exactly what he wanted (GPL'ing of Qt) and not once does he say thank you to trolltech," one Slashdot user wrote: "Instead, he switches from license bashing, to other forms of insults." Another satirically mocked Stallman's refusal to embrace the KDE community until it begged forgiveness:

Oh what a wicked generation of thieves and harlots. Repent now, and be saved. Accept the One True Way(tm).

Blessed are they who walk among the gnomes, for they will be Free(tm).

Blessed are those [who] change their licenses, for they will be forgiven.

Blessed are those who assign copyright to the FSF, for they will inherit the Kingdom of GNU(tm).

If you truly be followers of RMS [Richard Stallman], you must daily take up your soapbox and follow him.

—I PERENTHIANS 2:14–20.[40]

Despite such tensions, Trolltech's decision to release Qt as FOSS signaled an important victory for the FOSS community in one of its first major confrontations with a proprietary software company. The battle over the Qt license was not comparable in scale or in kind to the struggle against Microsoft that also erupted in the late 1990s. But it was an example of the FOSS community's ability to secure the leverage it needed to overcome proprietary software companies whose leaders chose to play by rules with which FOSS developers disagreed.

The Qt affair was a victory in particular for the "free software" segment of the FOSS community, which centered around GNU and Stallman. This faction's counterpart, the "open source" camp, considered the Q Public License to be a valid open source software license. But GNU developers were able to orchestrate enough support to prevail in their efforts to suppress use of any non-GPL-protected version of Qt. They sent a message that compromise with proprietary-software companies was not necessary.

As for the GNU desktop environment projects, the release of the Qt library under the GPL obviated the need to continue work on Harmony, whose developers declared the project "effectively dead" on August 4, 1999.[41] GNOME, however, remained under development. Version 1.0 of the desktop environment appeared in 1999, followed by the 2.0 version in June 2002. By the time of the 2.0 release, GNOME had evolved into a feature-rich interface that offered customizable themes, internationalization options, and accessibility support for users with impaired vision or reading abilities. It also supported a range of operating-system platforms, including not only systems based on GNU, Linux, or BSD but also proprietary alternatives, such as Solaris, HP-UX, and Unix itself.[42] Some third-party groups even attempted to port GNOME to Microsoft Windows in the early 2000s, with moderate success.[43]

### Office, Email, and Cross-Compatibility Applications

KDE, GNOME, and other desktop environments that emerged in the FOSS world around the turn of the millennium did much to make FOSS operating systems more pleasant and convenient platforms for users who preferred to work with graphical interfaces. Yet their effect was multiplied by the evolution of productivity applications, such as office suites and sophisticated email programs, which also greatly refined the FOSS user experience.

There were several attempts to bring modern office suites to the FOSS community at this time. One was GNOME Office, a short-lived package released in 2003 that failed to gain a large following. GNOME Office combined the GPL-licensed AbiWord word processor with database and spreadsheet software developed by GNOME programmers.[44] Another effort, discussed earlier in this chapter, was Corel's attempt to make WordPerfect available in a GNU/Linux environment.

Yet one office FOSS productivity platform fared better than these. By the mid-2000s, OpenOffice.org, which included a word processor, spreadsheet program, presentation software, and other tools, dominated the FOSS world. It also established a significant presence within the proprietary one.

OpenOffice.org originally was a proprietary office suite called StarOffice, which the German company Star Division began developing in 1985. In October 2000, Sun Microsystems acquired the StarOffice source code and released it to the public as an open source project named OpenOffice.org. The first stable version of the office suite appeared a year later.

By 2004, OpenOffice.org controlled 14 percent of the enterprise market, a significant feat for a project that faced entrenched competition from proprietary competitors like Microsoft Office. The achievement resulted, in part, from Sun's investment of significant sums of cash in OpenOffice.org development, which relied more heavily on paid programmers than did most other FOSS projects of the time.[45]

OpenOffice.org remained the most popular office suite for FOSS systems until 2010, when Oracle's acquisition of Sun fractured the OpenOffice.org development community. At that time, a team of programmers who were wary of Oracle's commitment to keeping the code free launched an alternative office suite, LibreOffice. LibreOffice was a *fork* of OpenOffice.org, meaning it was based originally on the OpenOffice.org code but was developed and distributed independently. In August 2011, OpenOffice.org came under the governance of an Apache license, which helped to allay concerns regarding the freedom of its code. Nonetheless, even though some GNU/Linux distributions continue to ship with OpenOffice.org today, LibreOffice has gained a wide following in recent years. It is now the default office suite on most of the flagship distributions.

Another productivity coup for FOSS users arrived with the 2001 release of Evolution 1.0, a free email client developed by a company named Ximian. Evolution was not the first FOSS email program, but it offered more features than the alternatives, as well as a rich graphical interface. Equally important was its support for email accounts hosted by Microsoft Exchange, a closed source email server that was widely used in the enterprise. Exchange support initially required the purchase of a proprietary plug-in for Evolution, but Novell's acquisition of Ximian in 2003 led to the release the next year of Evolution 2.0, which included Exchange support for free.[46] Attempts to port Evolution to Windows and Mac OS X in the mid-2000s saw little success, but Evolution remains a popular email client for GNU/Linux distributions today.[47]

FOSS users gained a second feature-rich email option with the release in summer 2003 of Mozilla Thunderbird. Although the first version of Thunderbird failed to gain a sizable following, the program saw wider adoption with the release of Thunderbird 1.5 in 2006. Evolution's tight integration with the GNOME desktop environment, along with KDE's reliance on native email clients of its own, positioned Thunderbird somewhat awkwardly within the application stack of GNU/Linux systems. Yet Thunderbird's support for multiple platforms, including Windows and Mac OS X as well as FOSS systems, ensured that the program would have enduring relevance as a leading open source email and collaboration solution. Mozilla's decision in

2012 to reduce its investment in Thunderbird slowed development significantly, but community programmers continue to work on it today.[48]

By the early 2000s, the introduction of programs such as Thunderbird, Evolution, and OpenOffice.org made FOSS operating systems much more viable productivity platforms for end users. But the lasting presence of the Wine project as a significant part of the FOSS landscape attests to the enduring demand of FOSS users for more applications than those natively available to run on free Unix-like systems. As noted above, Wine is a compatibility layer that allows operating systems based on the Linux kernel to execute programs compiled for Windows. As the project's acronymous name implies, Wine is not an emulator, meaning that it does not virtualize a Windows system or require Microsoft software to run. Instead, Wine implements reverse-engineered functional equivalents of Windows system calls within a Linux environment. This means that applications that run via Wine theoretically perform about as well as those running natively on Windows and can sometimes be even faster.

The ability to run Windows applications on a GNU/Linux distribution might appear to be advanced functionality, and Wine did not become commercially important until the launch of platforms like Corel Linux in the late 1990s. Yet Wine is nearly as old as the Linux kernel itself, and it was not an especially innovative idea at the time of its birth. In the early 1990s, when the operating system ecosystem was much more diverse than it is today, demand was high for solutions that allowed programs designed for one type of system to be run on a different platform. This was true not only in the Unix-like world but beyond. IBM also built a Windows binary-compatibility mode into its OS/2 operating system in the 1990s, for example.[49]

The development of Wine dates to June 1993, when some members of the early Linux programming community—who were inspired by a proprietary compatibility layer called Wabi that Sun had demonstrated in 1993 to run Windows programs on its Solaris operating system—began collaborating to develop a similar solution for Linux. Although Torvalds was not involved in Wine development, he was an early backer of the idea.[50] By July 1994, the Wine team had its own Usenet group—comp. emulators.ms-windows.wine. Development remained steady despite the introduction of 32-bit programs for Windows later in the 1990s, which made it more difficult to implement binary compatibility between Windows and Linux applications.

Beginning in 1998, Corel marketed a GNU/Linux distribution that offered Windows compatibility via a customized version of Wine as one of its headline features, as noted above. Lindows followed Wine development closely as well, for similar reasons.[51] Microsoft was sufficiently wary of Wine to note in an internal memorandum from 1998 (which is one of the Halloween documents discussed in detail in chapter 5) that "Linux advocates … are working on various emulators and function call impersonators" that could make it easier for users to migrate away from Windows.[52]

The Wine-related endeavors of Corel Linux and Lindows failed to garner much commercial following, and Microsoft ultimately had little to fear from Wine. Even so, development of the platform continued steadily throughout the early 2000s. In March 2002, in response to concerns that a commercial entity might appropriate their work, Wine developers voted to adopt the Lesser General Public License (LGPL) to govern the Wine code, which initially had been available under a less restrictive BSD-style license. (A subsequent fork of the project called Rewind, which retained the original licensing terms, fizzled due to lack of developer interest.) Three years later, Wine developers released the first "beta-quality" version of the platform, which they followed in 2008 with the introduction of Wine 1.0. The Wine compatibility layer remains a common feature of many GNU/Linux distributions today, enabling FOSS users to run applications such as Microsoft Office. Ports also have extended Wine's functionality to other Unix-like operating systems, including the Darwine project for Mac OS X.[53]

## FOSS AND THE WEB:
## APACHE, SAMBA, PHP, AND MYSQL

The rapid expansion of the scope and sophistication of the FOSS software suite in the 1990s and early 2000s—when GNU/Linux distributions proliferated and applications grew increasingly diverse—was a clear indicator of the viability of FOSS development models. However, many of these new platforms centered on desktop computing, where—despite endeavors like Corel Linux and Microsoft's occasional expression of concern over products such as Wine—they enjoyed relatively limited adoption. The saturation of the desktop market with proprietary software made this difficult ground for FOSS developers and companies to conquer.

The story was different in the rapidly expanding market for the software that powered Internet-connected servers. In the 1990s, the explosive importance of FOSS applications and operating systems for servers demonstrated FOSS's large-scale viability more clearly than ever. It also eventually placed the FOSS ecosystem squarely in Microsoft's sights, fomenting the "war" described in the following chapter.

In most ways, it is not surprising that FOSS enjoyed its greatest successes during the 1990s in the Internet market. Of all the technology sectors at the time, the Internet depended most closely on the principles and practices that were central to FOSS developers. The Internet was a decentralized, collaboratively maintained network. It functioned because open standards and protocols allowed computers that were built with different hardware and ran diverse operating systems to send email, files, Usenet posts, and eventually HTML-based Web pages to one another. "The Internet is, in many ways, the original Open Source venture," as one set of chroniclers wrote in 1999 in an essay about what they called the "Open Source Revolution."[54]

When discussing the 1990s, however, generalizations should not be made about openness in different parts of the Internet. Although open standards were important in the history of the Internet as a whole, in the early 1990s it was less clear that openness would prevail within the emerging segment of the Internet known as the World Wide Web. Tim Berners-Lee famously created the Web starting in 1990 at the CERN (Conseil Européene pour la Recherche Nucléaire or European Organization for Nuclear Research) lab in Switzerland by developing the hypertext markup language (HTML); a browser that could display documents in HTML format; a server that served HTML files; and the HTTP protocol, which allowed communication between servers and browsers. CERN distributed its Web software relatively freely, but there was little collaboration between the organization and the FOSS movement in the years when the Web was emerging.

By around 1994, according to CERN's official history, "the Free Software movement had become ... better known at CERN."[55] It was probably for that reason that CERN in 1993 released its Web software into the public domain. However, as Stallman emphasized in the early 1990s, public domain software was not at all the same as adopting free software because placing code in the public domain does not ensure that it will always remain open.[56] CERN later endorsed the GPL, but initially, its programmers did not consciously attempt to build the Web using software protected by the same licenses that governed Linux, GNU, and other major FOSS projects of the era.

In this sense, it was significant that the Apache Web server succeeded spectacularly in making FOSS code a key part of the Web in the mid-1990s. Apache's origins can be traced to a public-domain HTTP server called NCSA HTTPd, which was developed at the University of Illinois's National Center for Supercomputing Applications as an alternative to the Web server that Berners-Lee had created at CERN. NCSA HTTPd evolved into the most popular server for the Web in the early 1990s, but Netscape's poaching from the project of its lead developer, Rob McCool, in 1994 stunted its momentum. Administrators of websites around the world then found themselves modifying their versions of NCSA HTTPd locally to suit their various needs, leading to incompatible implementations of the software. "Uncertainty over the future of the NCSA httpd license" was also a concern, according to developers.[57]

Apache was born as an effort by a small group of webmasters to solve this problem. By February 1995, eight developers had begun collaborating as an organization they called the Apache Group to build a new Web server based on the NCSA HTTPd code base.[58] "Our goal," as one of the founding developers, Robert Thau, described it at the project's start, "is to produce a revised version of NCSA [HTTPd server version] 1.3 which has all the popular fixes in it directly, in order to have a supported server which actually meets our needs."[59] They relied on a series of patches to do that, giving rise to the server's name, Apache, which was a play on the phrase "a patchy server."[60] (Months into the project, Apache developers confronted concerns that the title might lack "political correctness," which they planned to address by vowing to change the name "if any authorized representative of the Apache Nation asks us" to do so. They also proposed offering Web hosting for "native-American-related pages" to mollify people who objected to the name.)[61]

By mid-March 1995—despite disagreements among the early developers regarding whether to announce their project publicly or keep the initiative to themselves—Apache had a logo and an official mission statement, which declared the group's intention to create the de facto server for the Web based on open standards:

> The Apache project has been organized in an attempt to answer some of the concerns regarding active development of a public domain HTTP server. The goal of this project is to provide a secure, efficient and extensible server which provides HTTP services in sync with the current HTTP standards.[62]

Over subsequent months, the development team overhauled the server's design and added new features, leading to the release in December of Apache 1.0. By early 1996, less than a year after the project's launch, Apache had become the world's leading Web server, according to usage statistics from Netcraft.com.[63]

The Apache Group introduced an important freely redistributable software product to the Web at a crucial moment. However, the launch of the Apache server did not ensure the influence on the Web of existing FOSS initiatives as much as it gave rise to a new faction within the FOSS community. As the Apache server and the other software that came under the Apache Group's umbrella grew in importance, the Apache community took steps to distinguish itself from other FOSS projects.

One of those distinctions centered on licensing. The Apache 1.0 license allowed free redistribution of the Apache server in source or binary form. It also permitted developers to create and distribute derivative works of the software so long as they clearly distinguished their work from the original product and retained the same licensing terms. However, the license did not compel or even encourage developers of derivative works to share their source code publicly, and it contained a clause that required all advertising materials for software packages that included Apache to give credit to the Apache project.[64]

For multiple reasons, these licensing terms sparked tense debate within the FOSS community. A key issue was the lack of protection for source code in the Apache 1.0 license, which did not meet the Free Software Foundation's standards. There were practical objections to the licensing, as well. By July 1995, the clause regarding advertising prompted complaints to the Apache developers from Yggdrasil Computing that the licensing terms made it unfeasible to include Apache in the company's GNU/Linux distribution. "I understand the intent behind the Apache license, but I feel that other licenses may do a better job of legally protecting the Apache Group and keep the price low," a Yggdrasil representative informed the Apache team. He suggested that "one license that would accomplish the same goal is the GNU General Public License, which also prevents proprietary extensions of software from becoming not free."[65]

The Apache programmers' reactions to such advice highlighted the wariness of some members of the FOSS community to accept aggressive free software licenses such as the GPL. As one developer wrote at the time:

> I'm no legal expert, and I don't even understand all of the GPL, but from other discussions, I get the feeling that it introduces many headaches, and that a [liberal] copyright would be a better idea. GPL wants to be the exclusive licensing agreement, so if you include any GPL source code in your product, everythign [sic] else that ships with it must also fall under GPL. Or something silly like that. That may become a factor if Apache incorporates code under different copyrights and distribution agreements.[66]

Another developer commented that if the Apache team were to adopt the GPL, then the move "would only cause us similar grief in the BSD camps" because supporters of permissive, BSD-style licensing terms like those of Apache would have been displeased to see the server come under the governance of the GPL.[67]

Ultimately, the Apache team opted to retain the original license for the duration of the 1990s. The developers modified the licensing terms in 2000, when they removed the burdensome language requiring advertising for Apache software to give the programmers credit. Instead, only the documentation supplied with GNU/Linux distributions and other software packages that contained Apache needed to mention the server's developers.[68] A further update in 2004, when the Apache 2.0 license was introduced, gained the approval of the Free Software Foundation, which deemed the new license compatible with the GPL.[69]

In addition to licensing differences, the Apache Group also adopted a somewhat different approach to code development in certain respects. Most significantly, it consciously avoided the "benevolent dictator" model that prevailed in some other major FOSS projects. As the organization's official history notes:

> Unlike other software development efforts done under an open source license, the Apache Web Server was not initiated by a single developer (for example, like the Linux Kernel, or the Perl/Python languages), but started as a diverse group of people that shared common interests and got to know each other by exchanging information, fixes and suggestions.[70]

For that reason, the Apache developers granted direct access to their source code repository to anyone whom they deemed—on the basis of suggestions or small code contributions submitted via email—deserving of becoming an official member of the project. This strategy was different from the carefully choreographed development methods of the GNU project. And although in a certain respect it resembled the "bazaar"-style approach of the Linux kernel, it was a more liberal version of this model. Many developers could propose changes to Linux, but their modifications required explicit approval from Torvalds before becoming part of the official Linux code. In contrast, the Apache approach allowed a broader group of developers to control the code directly.

The appeal of the Apache model for many FOSS developers was made clear by the rate at which the organization grew.

Although conceived initially only to develop a Web server, the Apache Group expanded its focus during the late 1990s to include a number of other projects. A 1996 proposal for creating an Apache Web browser never bore fruit, but by 1999, the organization was launching initiatives such as Jakarta, which develops several different programs based on the Java programming language. So-called because the project was conceived in a Cupertino, California, conference room of the same name, Jakarta was made possible by Sun's agreement to transfer Java code to the Apache developers "so that development of these technologies could take place in an open and collaborative way."[71]

The growth of the Apache Group's purview, combined with the commercial significance of its Web server software, spurred calls for the project to establish a more formal structure. In October 1998, the core Apache developers—who by that date totaled eighteen—met for the first time in person at the inaugural ApacheCon conference in San Francisco. They were joined by nearly five hundred other attendees, as well as representatives of IBM, Red Hat, and approximately ten other companies, which provided proof of Apache's importance for industry.[72] Several months later, in March 1999, the Apache Software Foundation was incorporated "to formally shepherd the development of the #1 Web server worldwide—the Apache HTTP Server Project—and other projects under the Apache umbrella."[73] With the incorporation, the loose collaboration that began in 1995 between programmers in the Apache Group became another major institutional presence within the FOSS community.

Again, that did not mean that Apache community leaders saw eye to eye on all issues with their counterparts in the Linux community, the Free Software Foundation, or other FOSS groups. The announcement of the Apache Software Foundation's incorporation quoted one of the founding board members, Brian Behlendorf, who said that part of the foundation's goal was to "do what we can to make the open-source development model really work," suggesting that the Apache supporters perceived flaws in the operations of other FOSS organizations. In documents related to an apparently abortive effort to write an official history of the Apache project in the early 2000s, Apache supporters similarly stated that "one of the most important roles that Apache plays, apart from being a damned fine product, is as a model of how [open source software] projects are *supposed* to work."[74]

Despite differences in licensing and development methodologies from other factions within the FOSS community, the Apache Software Foundation has thrived since its creation more than fifteen years ago. By 2016, it hosted nearly three hundred different FOSS projects.[75] Only a minority of them are directly related to Web server software. Instead, the list includes projects such as Hadoop, an important platform for processing big data, and Cassandra, a distributed database system that is designed for cloud computing infrastructure. As one of the most important centers of FOSS software development since the late 1990s, Apache has shown that the GPL is not the only way to license FOSS software effectively. It also highlights the rich diversity of FOSS ideologies and methodologies and makes the point that there is no single way to "do" FOSS.

The Samba project was another important component of the FOSS community's ability to compete against proprietary encroachment in the Internet market in the 1990s. Launched in late 1991 by a lone programmer, Andrew Tridgell, who reverse-engineered Digital Equipment Corporation's networking protocol for servers, Samba evolved by the late 1990s into a full-featured suite for integrating different types of operating systems across a computer network. This meant that companies using Windows and GNU/Linux-based computers concurrently could share data, printers, and resources between devices over the network. In the Samba developers' own words, the software made it "possible for many kinds of systems to share files that have been incompatible until now." As such, Samba, which was and remains licensed under the GPL, became an important tool for resisting efforts by proprietary software companies to "lock in" customers to their products by making them incompatible with third-party solutions.[76]

Samba was also significant as the first major FOSS project to confront challenges that arose from closed protocols rather than closed code. Unlike the developers of Linux and GNU, Tridgell and his team were not working to create new networking software that would replace proprietary solutions. Their goal was instead to discover the details of the protocol that defined how Windows computers exchanged information over the network, which was not publicly documented. Samba increased, rather than undercut, the market share of a proprietary networking protocol—albeit by offering a FOSS alternative to the closed source software that Microsoft distributed. In this way, the Samba developers showed that the FOSS community could mount effective responses to organizations that sought to limit the openness not only of code but also of protocols and application programming interfaces (APIs).

In a similar way, free programming languages designed for the Internet age helped to keep standards open as the Web grew increasingly important, just as the C language had done for Unix decades earlier by providing an easily portable coding framework. These included PHP, which was born in 1994 out of an effort by the Greenlandic programmer Rasmus Lerdorf to track visits to his personal website. (He accordingly named the language *Personal Home Page*, from which the shorthand *PHP* was derived.) In June 1995, Lerdorf made the source code for the PHP package, which by then had expanded to support database interaction and the creation of dynamic Web pages, publicly available.[77]

Initially, PHP worked only on Unix-like operating systems and had a relatively small following. In May 1998, a mere 1 percent of the world's Web servers deployed PHP. After an Israeli company took over PHP development from Lerdorf later that year, however, and released PHP 3.0 with additional features and support for Windows servers as well as Unix-like ones, the platform's popularity grew. At its peak, PHP 3.0 ran on about 10 percent of servers around the world. The release of PHP 4 two years later made the programming language yet more popular with Web developers, and PHP 5, which debuted in July 2004, enjoyed even wider deployment, thanks in part to the decision by Mark Zuckerberg to adopt it as the basis for building Facebook. PHP's popularity has declined somewhat in recent years, but it remains a vitally important FOSS language for Web programming today.[78]

Alongside PHP, the Perl and Python languages also emerged as favorites of the FOSS programming community in the 1990s. These languages, which date respectively to 1987 and 1989, emerged as the efforts of individual programmers to create elegant high-level programming languages tailored to the needs of hackers.[79] Although neither Perl nor Python caters to Web development in as direct and exclusive a way as PHP does, they nonetheless helped to drive innovation as FOSS developers defended their slice of the Internet market.

The last essential ingredient in ensuring FOSS's leading role as the Web grew was MySQL, a database for storing information. MySQL originated from database software called Unireg that Michael "Monty" Widenius developed in 1979 for a Swedish company named TcX. Starting in 1994, Widenius and David Axmark began work on an updated version of Unireg, called MySQL, which they released publicly in October 1996. The first public version of MySQL was available only in closed source form for the Solaris operating system, but a Linux release with full source code debuted in November 1996.[80]

MySQL's appearance filled a crucial niche in the burgeoning FOSS software stack for the Web. It provided a FOSS solution for storing data, which an Apache server could then retrieve and deliver over the Web in response to commands from PHP scripts. Running on top of a Linux-based server, Apache, MySQL, and PHP constituted what systems administrators came to call a LAMP stack. As a cost-free and customizable way to create the software infrastructure needed to serve Web pages or other applications, the LAMP stack had obvious value as the Web entered millions of homes and businesses in the late 1990s. The LAMP stack, which academic researchers in 2011 said "approaches the intellectual complexity of the Saturn V rocket," remains common today.[81]

## FOSS AND BUSINESS

The success of Apache, Samba, MySQL, and other Internet-related FOSS projects in the 1990s helped to diversify the FOSS community. It also raised the importance of FOSS within the business world to a new level. As earlier chapters of this book show, profitable commercial ventures related to FOSS were not new in the late 1990s. Companies such as Cygnus Solutions had been making money by selling support and other services related to free software for a decade by that time. And in the early 1990s, a variety of start-ups experimented with business models that centered on Linux. Yet by the end of the decade, a larger set of companies was turning profits by leveraging FOSS's newfound relevance within the Internet market as well as the many other FOSS applications and operating systems that had become available.

One of these companies was Red Hat, Inc. The business traced its origins to the humble ACC Corporation, an outfit owned by Bob Young that distributed software and documentation for Unix-like systems in the early 1990s. In January 1995, realizing that sales of Linux-based CDs were outpacing his documentation business, Young partnered with Marc Ewing, creator of the Red Hat GNU/Linux distribution, to found Red Hat Software, Inc. The new company intended to sell Red Hat GNU/Linux and other FOSS products through major retail outlets, such as CompUSA.[82]

For Red Hat, identifying a business goal proved easy, but developing a strategy for achieving it was harder. Young wrote in a 1999 essay on the company's history that he and Ewing initially considered several different types of industries to emulate as they pondered how to profit by distributing software that cost nothing in its original form. One of the industries they contemplated was the legal industry because one of its chief commodities, legal arguments, is free and exists in the public domain. Another idea was to mimic the automotive industry, where companies use an assembly process involving parts from hundreds of distributors to manufacture a vehicle and then guarantee the final product, just as Red Hat intended to do by combining an array of different FOSS programs to build its GNU/Linux distribution.[83]

Ultimately, however, Red Hat settled on the model of the commodity industries, including companies that sell such items as bottled water, soap, and ketchup. For these companies, success is based on marketing strategies that build strong brands, without which their products would have little value. "Ketchup is nothing more than flavored tomato paste," Young wrote: "Something that looks and tastes a lot like Heinz Ketchup can be made in your kitchen sink without so much as bending a copyright rule." Yet at the time of Young's writing, Heinz controlled 80 percent of the ketchup market, and almost no one made ketchup at home.[84] One reason that this is so is that buying ketchup from Heinz is more convenient than making it. But more important, Heinz has defined how ketchup should look, smell, and taste, according to Young. In the same way, Red Hat sought from its early days "to offer convenience, to offer quality, and most importantly to help define, in the minds of our customers, what an operating system can be," in Young's words.[85]

Red Hat faced stiff competition in that regard from Microsoft, which had a two-decade head start in the quest to define the type of operating system users thought they should want. As a result, Red Hat set out to distinguish its products by emphasizing their ability to give users control over software. In that respect, the company transformed a core tenet of the FOSS movement—Stallman's aggressive emphasis on user freedom as opposed to development efficiencies—into a marketing pitch. The company thus showed that, even though some FOSS supporters found rhetoric related to freedom overly bombastic, the Free Software Foundation's message was not incompatible with commercial endeavors. Red Hat also proved that selling support services related to free software, as Cygnus Solutions had done, was not the only way to make money in the FOSS market. Customers would also pay for code if it came compiled and packaged in a compelling way that added value, even though they could obtain the constituent parts elsewhere for free.

With this approach, Red Hat flourished. On August 11, 1999, it became the first FOSS business to go public. In November of the same year, the company merged with Cygnus Solutions, combining the latter's support services for FOSS with Red Hat's enterprise-grade GNU/Linux distribution. Although the FOSS commercial space has grown more diverse since that time, Red Hat remains the flagship company in this market today and is among the top corporate contributors to Linux kernel development.

The other major FOSS company to go public in the late 1990s, VA Linux Systems, fared less well. Founded in 1993, VA Linux attempted to pioneer yet another FOSS-based business strategy. It sold computers with GNU/Linux preinstalled, aiming to compete with the likes of Dell.[86] The company expanded rapidly and was making $100 million in annual sales by 1998. In the same year, it received capital investments totaling $5.4 million from Intel and Sequoia Capital. The next year, an additional $25 million in funding arrived from an assortment of other backers.

On December 9, 1999, to great anticipation, VA Linux became a publicly traded company, operating under the stock symbol LNUX. Priced initially at $30 per share, the company's stock gained 698 percent in its first day of trading, the highest rise in NASDAQ history. But the momentum of VA Linux did not endure. A year after the company's stock began trading publicly, LNUX had declined in value to $8.49 per share, about 3.3 percent of its initial high.

To most observers at the time, the astounding rise and fall of VA Linux reflected the wild nature of the broader Internet "dot-com bubble" of the late 1990s, when overvaluation of technology companies was routine. The *New York Times*, which summarized the company's initial public offering with the headline "A Tiny Company with Dim Prospects Goes Public with a Bang," was pessimistic about VA Linux's outlook even before the stock flopped.[87]

Yet VA Linux's failed public launch was not the result of the dot-com bust alone. Its fate was also attributable to its attempt to compete in the desktop computer market, where it fared no better against entrenched proprietary competitors than ventures like Corel Linux did in the same period. In addition, VA Linux suffered from a lack of strong contradistinction against other desktop and server hardware manufacturers at the time. Selling computers with Linux and GNU software preinstalled was not a unique value proposition.

Despite the bursting of its initial bubble, VA Linux survives today as a testament to the commercial viability of media related to FOSS software. In early 2000, the company began shifting its focus away from selling computer hardware after it acquired Andover.net, which owned a variety of news outlets that catered to the programmer crowd, including Slashdot. The next year, VA Linux changed its name to VA Software and excised itself from hardware sales entirely. After a series of subsequent mergers, the company today operates as Geeknet, Inc., and continues to focus on websites and other media offerings for the FOSS niche.[88]

It was not only startups like VA Linux and Red Hat that fueled the explosion of FOSS commercial activity in the late 1990s. Taking note of the investor enthusiasm with which these companies began trading publicly, as well as figures such as the one showing that businesses installed 750,000 Linux-based servers during 1998 alone, established companies took a keen interest in the Linux kernel, the Apache Web server, and GNU software during this period.[89] Of these, the most significant by far was IBM.

IBM began investing in the FOSS space in June 1998, when it announced that it would sell and support the Apache Web server.[90] From there, the company's interest in Linux grew as follows, according to Big Blue:

> By the late 1990s, Linux was almost good enough to run high-volume servers—and it was free. Sam Palmisano, at the time an IBM senior vice president, came back from a global tour of Internet companies and reported that he kept hearing all about Linux from young programmers. As IBM's head of internet strategy, Irving Wladawksy-Berger similarly kept hearing about Linux. IBM commissioned an internal study, which turned into a plan to use Linux to deliver innovation in new ways and create a new force of openness, quality, performance and security. Wladawsky-Berger and Palmisano supported this plan and helped convince then-CEO Lou Gerstner to make a big bet on the forces of open innovation. Adopting Linux as an IBM open operating system looked like a

gigantic, risky, counter-culture bet. But "Linux perfectly fit what we needed," Wladawsky-Berger said later. Over time, it became obvious that it wasn't a counter-culture bet at all, but a leadership step towards the mainstream culture of the future.[91]

In January 2000, IBM made its endorsement of Linux public. At the end of that year, it took the further step of announcing that it would invest $1 billion to support the development of the kernel and related projects. IBM's chief executive at the time, Louis Gerstner, said the company viewed Linux as "a major force in IT and moving IT to the next generation of the Internet business," particularly because he believed that open, standards-based systems were the future. "The movement to standards-based computing is so inexorable, I believe Sun—and EMC and Microsoft for that matter—is running the last big proprietary play we'll see in this industry for a good long while," Gerstner said.[92]

As IBM's CEO indicated, the company's embrace of FOSS, which Torvalds called "Linux's biggest coup," was about much more than altruism.[93] By supporting the development of GNU/Linux systems, IBM aimed to ensure that Microsoft, which by the late 1990s had cornered most other sectors of the operating system market, would not monopolize software for server platforms. If it did, IBM or its customers would likely have faced steep prices when deploying the server hardware that IBM sold. For IBM, the combination of Linux with GNU programs and Web software such as Apache's to build freely redistributable server operating systems promised to prevent that eventuality.

IBM's billion-dollar Linux investment infused enormous cash into the FOSS community, augmenting the momentum of FOSS development even further.[94] And beyond cash, IBM's investment in Linux lent crucial credibility to the FOSS ecosystem by showing that multinational corporations believed that code that cost nothing was worth billions of dollars.

While IBM was pouring vast sums into FOSS development, another long-established technology company, Apple, was making FOSS investments of its own as it built a new operating system on a FOSS foundation. The OS X operating system, which Apple introduced in 2001 as a major reworking of the software platform it included on the computers it sold, originated from the Unix-like OPENSTEP operating system that the company NeXT had developed under the direction of Steve Jobs during his hiatus from Apple from 1985 to 1997. OS X was based on the Mach microkernel (the same one that the GNU developers experimented with, as chapter 2 notes) as well as components drawn from the BSD operating system (such as the networking stack). Because all of this software was available under permissive licenses that allowed free redistribution, Apple was able to integrate the code into OS X without legal or financial constraints.

Apple's inclusion of Mach and BSD code in OS X ultimately brought FOSS to the computers of millions of people who might not otherwise have used it, but the development did not sit well with leading members of the FOSS community, which has tended historically to have a tense relationship with Apple. In part, the animosity predated the development of OS X by nearly a decade. As chapter 2 notes, the Free Software Foundation encouraged supporters to boycott Apple products between 1988 and 1995 as a result of the company's efforts to sue competitors for infringing on the "look and feel" of Apple products.[95]

Animosity toward Apple was also evident in Torvalds's 2001 autobiography, where he lashed out against a company that, perhaps more than any of its peers, has built a thriving business selling highly proprietary software with a FOSS core. Torvalds describes a meeting with Jobs in the late 1990s, when OS X remained under development: "Basically, Jobs started off by trying to tell me that on the desktop there were just two players, Microsoft and Apple, and that he thought that the best thing I could do for Linux was to get in bed with Apple and try to get the open source people behind Mac OS X."[96] The creator of Linux was not at all receptive to this idea, partly because he believed that, from a technical perspective, the OS X kernel was "a piece of crap. It contains all the design mistakes you can make, and managed to even make up a few of its own."[97]

Philosophical disagreements over kernel design were only part of the issue. Torvalds also found Jobs to be out of touch with the purpose and potential of Linux and FOSS generally. "Jobs made a big point of the fact that Mach's low-level kernel is open source," Torvalds writes," yet "he sort of played down the flaw in the setup: Who cares if the basic operating system, the real low-core stuff, is open source if you then have the Mac layer on top, which is not open source?" For Torvalds, Jobs's

> main argument was that if I wanted to get the desktop market I should come join forces with Apple. My reaction was: Why should I care? Why would I be interested in the Apple story? I didn't think there was anything interesting in Apple. And my goal in life was not to take over the desktop market.[98]

Torvalds also disdained Jobs's self-flattering approach, writing that "He just basically took it for granted that I would be interested. He was clueless, unable to imagine that there could be entire segments of the human race who weren't the least bit concerned about increasing the Mac's market share. I think he was truly surprised at how little I cared about how big a market the Mac had—or how big a market Microsoft has."[99]

Stallman was similarly critical of Jobs and Apple long after the "look and feel" lawsuit battles of the late 1980s and early 1990s ended. Following Jobs's death in 2011, Stallman, misquoting the statement of a former mayor of Chicago in response to the passing of a political opponent, wrote on his personal website:

> "I'm not glad he's dead, but I'm glad he's gone." Nobody deserves to have to die—not Jobs, not Mr. Bill, not even people guilty of bigger evils than theirs. But we all deserve the end of Jobs' malign influence on people's computing.[100]

In a subsequent post, Stallman clarified the misquotation of the Chicago mayor but remained unapologetic about his comments on Jobs's efforts

> to make general-purpose computers with digital handcuffs more controlling and unjust than ever before. He designed them to refuse even to let users install their own choice of applications—and installing free (freedom-respecting) applications is entirely forbidden. He even tried to make it illegal to install software not approved by Apple.[101]

In 2013, Stallman reaffirmed his position on Jobs and Apple once again. He told Internet users that, as crude as his comments after Jobs's death may have been, the issue was one of "substantive good and evil," which transcended concerns of political correctness.[102]

Despite strongly worded criticisms of Apple such as these, the tensions between the company and the FOSS community in the late 1990s could not compare to the existential struggle that erupted between FOSS and Microsoft. That conflict, along with battles within the FOSS community that comprised the other part of the FOSS "revolutionary wars," is the subject of the next chapter.

# 5  THE FOSS REVOLUTIONARY WARS
## Free Software, Open Source, and Microsoft

WARFARE FREQUENTLY ACCOMPANIES revolutions, and the free and open source software (FOSS) revolution was no exception. In the late 1990s and early 2000s, the FOSS revolutionaries found themselves fighting not one but two wars born out of the momentous changes their activities had brought about in the world.

They waged the first of those conflicts among themselves. As hackers with divergent interpretations of what free software—or *open source software*, to use the term coined by one of these groups in 1998—should be and do vied for control of the FOSS universe, an ongoing rift developed within hackerdom.

Meanwhile, the FOSS community as a whole found itself waging a different struggle against external enemies. Alarmed by the way FOSS was revolutionizing the computing industry in the late 1990s, proprietary software companies, with Microsoft chief among them, poured huge amounts of resources into an effort to destabilize the FOSS community and stymie FOSS software's surge into the business world. The FOSS revolutionaries confronted this challenge at the same time that they battled among themselves.

This chapter explores both of these revolutionary wars, their influence on the technical and political evolution of FOSS, and their outcomes. It also explains why the winning factions—the open source camp in the case of the FOSS civil war and the FOSS community in the war against Microsoft—triumphed.

## THE FOSS CIVIL WAR

Stallman and Torvalds were in agreement regarding Apple, as the end of the previous chapter notes, but by the late 1990s, their views on other important issues had grown much less harmonious. By that time, these programmers had emerged, willingly or not, as figureheads for competing factions within the FOSS community. These two groups—the "free software" camp (led by Stallman) and its "open source" adversary (associated with Linux and Torvalds)—found themselves battling for control over the meaning of the FOSS revolution and software freedom. This struggle became the FOSS civil war.

The FOSS civil war of the late 1990s was not completely novel. As previous chapters show, the free software community, like any large, voluntary group, had always lacked total consensus on various political and technical issues. In the 1980s and early 1990s, for instance, the GNU project and the BSD developers took very different approaches to licensing. Yet during that period, these developers got along well enough to continue collaborating with one another.

By the late 1990s, however, differences between the free software and open source camps had grown so irreconcilable that each group began excluding the other from its plans and activities. The factions also questioned the legitimacy of competing interpretations of the meaning of software freedom.

Previous attempts to explain the unprecedented tensions that affected the FOSS community in the late 1990s attribute them to a shift in the predominant hacker mores. That interpretation emerges from Raymond's essay "Homesteading the Noosphere," first published in 1998 and updated several times through 2000, in which he examines different strands of hacker culture. Raymond contends that "historically, the most visible and best-organized part of the hacker culture," which he identifies with Stallman and the Free Software Foundation, "has been both very zealous and very anticommercial." Only after "the Linux explosion of early 1993–1994," he writes, did a more pragmatic group of hackers establish "a real power base," which centered on Linux development.[1]

Painted in broad strokes, the historical narrative aligns with Raymond's suggestion that the center of gravity within the hacker community was shifting toward "pragmatism" by the mid-1990s. Yet a closer analysis reveals a more complex picture in several respects. For one, as chapter 2 shows, it is inaccurate to describe Stallman and GNU as unpragmatic or anticommercial, especially in their early years. Stallman and his cohorts were firm in their beliefs but did not oppose commercial endeavors related to free software. On the contrary, they helped to pioneer them. Nor were they unwilling to make pragmatic compromises, as they did, for instance, when scaling back the stringency of the GPL licensing terms in the late 1980s.

In addition, Raymond's depiction of Linux hackers as more pragmatic, less zealous, and less anticommercial than their GNU counterparts is a messy one. Those descriptors fit some of the Linux developers at certain points in the kernel's history. But developers like those in the Debian project, which was born out of a fear that the Linux ecosystem was becoming too commercial and was consequently losing its value for hackers, do not fall cleanly into Raymond's interpretation. At the same time, Torvalds himself, as the previous chapter shows, was doggedly anticommercial when he first released the Linux kernel, an attitude highlighted by the no-money clause of the original Linux license. This further complicates the notion of a clear distinction between the ways that the Linux crowd and Stallman and his cohorts thought about software and hacker values.

Finally, although hacker culture had evolved by the 1990s from what it had been decades earlier, the core hacker values endured. True, a singular, universal hacker ethic remained just as elusive in the 1990s as it was in the 1950s and 1960s. Yet as Steven Mizrach shows in a comparison of hacker literature from different time periods, most hackers in the 1990s continued to embrace a basic set of principles that included the imperatives of sharing and open communication.[2] Mizrach also demonstrates a clear continuity between these values in the 1990s and their origins among 1950s- and 1960s-era hackers. There was no fundamental shift in the type of hacker who stood at the fore of the community between the 1960s and the 1990s.

For all of these reasons, the FOSS civil war of the late 1990s cannot be described as a struggle between pragmatic and unpragmatic hackers or between different generations of hackers or groups that endorsed wholly opposite forms of the hacker ethic. The conflict centered instead on defining the meaning, goals, and limits of the FOSS revolution, which hackers had never clarified despite having spent many years working to advance free software projects.

Ambiguity regarding the nature and aims of the free software revolution had been embedded within the movement since its origins. As chapter 2 notes, when Stallman announced the GNU project in September 1983, he failed to specify what he had in mind when he used the word *free* to describe the Unix-like operating system he envisioned.[3] Readers found it easy to assume that Stallman meant that he would give away the software at no cost. The "GNU Manifesto," which Stallman published in 1985 and updated a few times in subsequent years, emphasized the importance of sharing source code more than the 1983 announcement had done, but the primary focus of the "Manifesto" remained on producing cost-free software. Stallman later admitted in comments on the document that his "wording was careless" and that when he published the text, he "had not yet clearly separated the issue of price from that of freedom."[4]

Yet even if Stallman eventually became better at articulating how, for him, ensuring the freedom of the user by distributing source code was the chief goal of free software development, the persistence during the GNU project's early years of ambiguity regarding what "free software" meant birthed a confusion that Stallman's later rhetoric could not fully erase. This confusion laid the foundation for different groups of hackers in the 1990s to stake competing claims regarding what free software should entail.

One of the earliest examples of this conflict involved debate over what to call GNU/Linux distributions. As chapter 3 explains, the common practice of referring to operating systems based on the Linux kernel and GNU software simply as "Linux" spawned a vociferous campaign by Stallman to promote the "GNU/Linux" nomenclature to emphasize what he called the "ethical importance" of free software.[5] That campaign was largely a failure, and even today, the term *GNU/Linux* rarely appears outside of circles of programmers who strongly support the Free Software Foundation. (This book uses the term *GNU/ Linux* out of a concern for balance and not as an endorsement of a particular FOSS ideology or camp.)

In 1999, one typical free software user suggested that Stallman lost the GNU/Linux battle because the term was simply too awkward to gain wide acceptance. "I recognize it as GNU/ Linux but I don't call it GNU/Linux because I'm lazy," he said.[6] Yet the notion of laziness only hints at the real issue at stake. For many hackers who opposed Stallman, the chief problem was not that the GNU/Linux name and other initiatives of the Free Software Foundation demanded too much of users but rather that these initiatives were not sufficiently utilitarian to serve what many people deemed to be the true goals of the FOSS revolution—which, for them, was about making coding more efficient, not fulfilling an ethical obligation.

The varying values that different hackers placed on ethics and utility was also evident in a major disagreement during the 1990s over the pace of development of the GNU C Library (glibc). A group of developers working on Linux felt that updates to the library were being issued too slowly to support the rapid pace of Linux development, so they forked glibc by launching their own version of the library, which they developed independently of the GNU version of the software.[7] The forking initiative proceeded for several years until GNU's release of a new version of glibc in January 1997 prompted developers of the competing implementation to embrace GNU's release once again. Even then, tensions remained bitter enough that

Stallman refused to integrate code from the fork back into the updated version of glibc, citing concerns over the ambiguity of authorship of the fork.[8]

At its heart, the glibc fork resulted from dissent regarding how utilitarian free software should be. The priority for GNU, which was still operating in cathedral-style development mode, was having infrequent, feature-rich, well-tested releases of the library because the GNU project's ultimate goal was creating an elegant free software operating system that would benefit users above all. In contrast, the programmers who forked glibc wanted an implementation that catered to the needs of the Linux kernel development community and was released early and often enough to sustain quick advances in Linux coding.

Disputes over whether utilitarian advantage should drive the big-picture vision for the FOSS movement escalated as the 1990s progressed. Matters came to a head in 1998, when Netscape announced that it would open-source its Web browser (a decision that led to the creation of Mozilla, which is discussed later in this chapter). The news prompted the publisher Tim O'Reilly, a FOSS supporter, to convene a meeting in April of what Raymond called "a select group of the most influential people in the Open Source community." Among them were Raymond, Torvalds, representatives of Mozilla and Apache, the creators of the Perl and Python programming languages, and several others.[9] Not present was Stallman, whom O'Reilly excluded because he deemed him "inflexible and unwilling to engage in dialogue."[10] Although O'Reilly invited other representatives of the GNU project, John Gilmore and Michael Tiemann, to the summit in Stallman's stead, the decision not to invite the father of the FOSS movement irked some hackers.

It also prompted Bruce Perens, of Debian fame, to refuse to attend in protest.[11]

The invitee list for the April 1998 meeting divided the hacker community into opposing factions, and the policies that emerged from the event inflamed the situation further. At the meeting, by a vote of nine to six, hackers officially endorsed the term *open source* as an alternative to *free software*, an idea Raymond had first suggested in February 1998.[12] The meeting attendees intended that the new terminology—which they approved after considering a variety of other ideas that included *freeware*, *sourceware*, and *freed software*—would serve the utilitarian purpose of convincing "the corporate world to adopt our way [of software development] for economic, self-interested, non-ideological reasons," according to a report Raymond published following the summit.

For Raymond and others who supported the *open source* nomenclature, catering to business interests was merely a means toward an end. Like Stallman, they also viewed the protection of hacker culture as the most important thing. Raymond even wrote that he deemed the open source movement to be necessary because it was the only way to liberate hackers from the "ghetto" in which they had dwelt for two decades by the 1990s, when they found themselves "walled in by a vast and intangible barrier of mainsteam prejudice" against hacker culture.[13]

All major figures in the FOSS community agreed on the importance of saving hacker culture. What separated them was that Raymond and his supporters viewed utilitarianism as the best means of achieving the ends they sought, while Stallman remained firm in his belief that abandoning the rhetoric of freedom would undermine the goal of protecting hacker culture.

To sustain their movement, the hackers in the open source camp also promoted a document written primarily by Perens called the "Open Source Definition," which defined what constituted an open source software license and differentiated open source from free software. In addition, the Open Source Initiative, an organization that had launched earlier in 1998 amid the excitement of Netscape's Mozilla announcement, ensured for the open source faction the same sort of institutional backing that the free software community received from the Free Software Foundation.

Predictably, Stallman and hackers close to him reacted with scorn to the open source campaign. At the 1999 LinuxWorld Convention and Expo, Stallman called the open source endeavor a "half measure" that doomed the free software movement from ever fulfilling its purpose of ensuring users' freedom.[14] Similarly, he described Torvalds in an interview in March 1999 as "basically an engineer" who "likes free software, but isn't concerned with issues of freedom." He added that, because of the open source campaign, "people are no longer exposed to the philosophical views of the GNU project," suggesting that GNU's high-minded vision was being compromised toward utilitarian ends.[15]

Stallman was not alone among prominent hackers in expressing such views. Perens—despite having written the "Open Source Definition" and straddling perhaps better than anyone else at the time the chasm that was dividing the FOSS community—resigned from the Open Source Initiative in 1999 because he believed that it had grown too distant from the Free Software Foundation.[16]

Raymond and other members of the open source camp fought back against the free software faction. They suggested that the Free Software Foundation's message failed to resonate with individuals who were not hackers, especially those within the business community. Raymond warned that Stallman's rhetoric "confuses and repels most people."[17] Torvalds, despite writing that Stallman "deserves a monument in his honor for giving birth to the GPL," nonetheless lamented that he "sees everything in black and white. And that creates unnecessary political divisions. He never understands the viewpoint of anybody else. If he were into religion, you would call him a religious fanatic."[18]

Part of the reasoning behind such criticisms of Stallman was the notion that his radical rhetoric stemmed from deepseated opposition to commercialism. That is why, for example, it was only after protests by Stallman that Raymond revised his "Homesteading the Noosphere" essay to register Stallman's "assertion that he is not anticommercial." Initially, Raymond argued in the text that Stallman was the leader of the part of the hacker community that was most anticommercial.[19]

In reality, as this chapter has already noted, such depictions of Stallman do not stand up to the evidence. Stallman and the Free Software Foundation disagreed with the open source faction's contention that making the FOSS movement as palatable as possible to business leaders should be a primary goal. But they did not oppose making money with FOSS as long as profit did not come at the expense of other priorities.

Portrayals of Stallman as antibusiness were also inconsistent with the image he tried to paint of himself, even though choices he began making in the mid-1990s made it hard for much of the business community to take him seriously. In 1996, for example, he invented a new persona for himself that he adopted at public events. He called himself St. Ignucius, dressed up in mock-religious garb with a paper halo on his head, and proclaimed himself a saint of the "Church of Emacs." Yet such activities, which Stallman has said represented an effort to poke fun at himself and to introduce some lightheartedness to the free software community, were not anticommercial or necessarily alienating.[20] They were just strange. Meanwhile, Stallman was keen to insist in the late 1990s that "I'm not against commercial anything," as he told one reporter.[21]

Ultimately, such statements exerted little influence on the perception of Stallman and the Free Software Foundation that has endured since the 1990s. Neither side completely won or lost the FOSS civil war. Both factions have continued to operate smoothly, though with limited mutual cooperation, since the 1990s. Nonetheless, the rift between them that emerged at that time has never fully closed. Stallman continues to denounce the *open source* terminology today. He told me, for example, that "it is a mistake to cite my statements as an example of the thought of open source, just as it would be a mistake to cite [Franklin Delano Roosevelt] as an example of conservatism. I am not a supporter of open source; what I stand for is free software."[22] And in contrast to figures like Torvalds, whom *Reader's Digest* named "European of the year" in 2001, Stallman has remained relatively unknown outside of technical communities.[23] The GNU camp did not lose the FOSS civil war, but it emerged from it wielding considerably less influence over the way FOSS code was developed and distributed than it had in the 1980s and earlier 1990s.

## THE WAR AGAINST MICROSOFT

Conflict within the FOSS community did nothing to calm the hawkishness of proprietary software companies, which by the late 1990s had grown increasingly wary of FOSS's expanding role in the business world. Attacks from these companies, especially Microsoft, fueled another war, which pitted FOSS advocates against external enemies as the FOSS revolution continued.

### Prelude to War: The Birth of Mozilla

The long-term causes of Microsoft's anti-FOSS campaign lay in developments that occurred over the decade following Linux's founding. Those happenings, described in the previous chapter, involved the introduction of new FOSS products, FOSS's endorsement by deep-pocketed companies such as IBM, and the launch of successful FOSS start-ups. Yet perhaps the most important trigger of Microsoft's war against FOSS was the birth of the Mozilla Web browser, which became an open source project in January 1998.

Unlike most of the other major FOSS projects of the 1990s, the Mozilla browser was conceived of as a business strategy. The developers of its progenitor—the proprietary, closed source Communicator browser owned by Netscape—evinced no deep-seated ideological commitment to the FOSS movement. The company was in the final stages of its losing war against Microsoft's Internet Explorer browser and hoped that opening the Communicator code to third-party contributions and distributing the browser free of charge would help to regain users and reduce development costs. Those goals were clear from the company's announcement of the change, which said it aimed "to accelerate development and free distribution by Netscape of future high-quality versions of Netscape Communicator to business customers and individuals, further seeding the market for Netscape's enterprise solutions."[24]

Yet because Netscape management prominently and publicly justified the open sourcing of Communicator with reference to what had become a canonical text of the FOSS community, Raymond's "The Cathedral and the Bazaar," Netscape's move served as a key endorsement of FOSS by a well-known technology company. The fact that the company had previously shown no interest in FOSS made the decision to open-source Communicator all the more striking. For these reasons, Mozilla represented a threat that gained Microsoft's attention in a serious and pressing way.

Netscape executives' consideration of the merits of open-sourcing Communicator dated to a whitepaper written by one of the company's employees, Frank Hecker. Hecker referenced Raymond's ideas regarding the pragmatic benefits of "bazaar"-style open source development models. Raymond's articulation of the strategy was new at the time, having begun circulating within hacker circles only in May 1997, when Raymond presented a paper at the Linux Kongress conference that became the basis for "The Cathedral and the Bazaar." (The essay was not formally published until August 1999, although versions of it appeared online before that time.)[25]

As Raymond noted, however, the conference paper and essay merely formalized the description of what some FOSS developers, particularly those working on Linux, had been doing for years by that point. "What I saw around me," Raymond wrote of the time when he began participating in Linux development, "was a community that had evolved the most effective software-development method ever and didn't know it! That is, an effective practice had evolved as a set of customs, transmitted by imitation and example."[26]

According to Raymond, Linux developers had accidentally discovered the principle that he called Linus's law, in honor of Torvalds. He defined Linus's law as follows: "Given enough eyeballs, all bugs are shallow." This meant that when many developers—and, ideally, users with programming skills—become involved in a software project, it becomes easier to identify and fix flaws and implement new functionality. That is why Raymond analogized Linux development to a bazaar, where many people interact rapidly and constantly with no central authority overseeing them. The bazaar in this sense operates quite differently from the construction of a cathedral, which a comparatively small team of builders slowly and steadily erects over a long period of time, with few opportunities to deviate from the central plan.

For a generation of programmers who had grown up following Fred Brooks's mantra that adding more developers to a project leads only to greater complexity and diminishing returns on labor—a principle that, as chapter 2 notes, was just as prevalent within the GNU developer community as it was within proprietary software companies—Raymond's interpretation of the merits of bazaar-style development was an inspiring innovation. For a company like Netscape, it also represented a logic that profit-oriented business managers could appreciate much more than the Free Software Foundation's rhetoric about morality and user freedom. Like others in the open source camp, Raymond emphasized the utilitarian efficiencies of bazaar-style development rather than ideological issues.

Spurred on by Raymond's arguments, the familiarity of many of Netscape's engineers with open source projects, and their recognition that open, community-developed languages such as HTML had been crucial to the company's previous successes, Netscape executives announced on January 23, 1998, that they would release the source code of Communicator publicly and invite third-party developers to contribute to the program.[27] Shortly thereafter, they set March 31 as a target date for making the code public. In the meantime they worked to excise proprietary sections of code from the Communicator code base—which developers inside the company called Mozilla, the name by which the public eventually came to know the browser—and perform other tasks necessary for publishing the code on the Internet.[28]

Netscape employees also spent the late winter and early spring of 1998 working to identify the best license to use for their browser after it became open source. After considering various existing licenses and consulting with Raymond, Torvalds, and O'Reilly, Netscape executives deemed BSD-style licenses to be too permissive. They found the GPL to be "untenable for commercial software developers" because it did not allow third parties to integrate their projects with Mozilla unless they also accepted the GPL. The Netscape team consequently opted to write a new, original license, the Netscape Public License, which it posted in draft form on the Internet on March 5, 1998, in order to seek feedback from the community. That was a novel decision because previous FOSS licenses had not been developed in a public way.[29] For a project embracing the bazaar mode of programming, however, asking for the community's help in writing a license made sense.

Reaching community consensus on licensing terms for Mozilla proved to be harder than expected. This challenge prompted the Netscape team to create an additional license, the Mozilla Public License. The Netscape Public License governed the original Mozilla code derived from Communicator, and the Mozilla Public License applied to contributions that developers made to the code base after Mozilla's birth. With this compromise, Netscape developers were able to protect the company's interests using the license they wrote themselves while also offering contributors a middle ground between BSD- and GPL-style licensing terms in the form of the Mozilla Public License. That decision pleased many programmers in the FOSS community.[30]

Completing their work just in time for the March 31 deadline, Netscape developers released the full source code for the Mozilla browser, which totaled 1.5 million lines, on that day. Some leading figures in the FOSS community celebrated this success with vibrant enthusiasm. As Torvalds recalled, this was especially true of Raymond, who "took it really personally" and assigned himself much of the credit for the process that resulted in Mozilla's debut.[31] Torvalds himself "thought it was wonderful that Netscape" open-sourced its browser, but in contrast to Raymond he "didn't view it as a personal achievement."[32]

Despite the support of FOSS leaders for Mozilla, the prospects of the open source browser seemed uncertain for some time following its release. Jamie Zawinski, one of the project's founders and its most important public face, resigned a year after the launch, citing mismanagement of the project and lost opportunities. Meanwhile, fewer third-party developers signed on to donate their time and expertise to Mozilla development than Netscape had hoped for. And Internet Explorer continued to dominate the market, leading some observers to question whether open-sourcing the Communicator code would prove an effective business strategy after all.[33] In 1999, even Raymond admitted that, as of that time, the project had proved to be "only a qualified success."[34]

Not until 2003, when Netscape (as the result of decisions by the leadership of its parent company, AOL) sharply reduced its support for Mozilla, did the project gain greater momentum. The news that Netscape would no longer fund significant browser development prompted the reorganization of the Mozilla project, which thereafter began developing a suite of applications rather than just the browser and received greater support from the community.

Although the Mozilla project's performance was lackluster during its first years, it served as a symbol of the FOSS community's strength as much as an embarrassment. As Raymond noted, Linux and other FOSS platforms continued to flourish and grow steadily in commercial importance even as Mozilla flagged. That success highlighted FOSS's evolution into a movement large and broad enough to sustain a major setback and keep moving forward.[35] It is easy to imagine a single crisis having demolished the free software community in earlier years. For example, had Linux not appeared at just the right moment to provide a GPL-licensed kernel for GNU to use, the problems with the Hurd may have fatally undermined the work of Stallman and his collaborators. By the late 1990s, however, the FOSS community possessed sufficient strength to work through setbacks.

### The Halloween Documents

The community was also strong enough to confront effectively the various attacks that Microsoft mounted against it in the wake of Mozilla's launch. These campaigns were based primarily on two main susceptibilities that affected FOSS software and the FOSS community. One of them was a perceived lack of usability in GNU/Linux distributions and other FOSS products, which was a perennial complaint among many computer users. In one typical example of such criticism, a Usenet reader in February 1999 published a post titled "Linux Sucks!! Long Live Windows," in which he cautioned, "I would avoid Linux like the plague. It may be good but it is by no means easy to install or remove from a hard drive. That's what the [sic] DON'T tell you when you try it out and as it is distributed free the author is not liable for your losses." Many readers agreed and were not persuaded by arguments that the lack of user-friendliness in Linux-based operating systems stemmed primarily from Linux's being newer than Windows.[36]

Hackers also recognized usability as a problem in much of the software they had built. They particularly lamented the paucity of good graphical user interfaces. In 1999, for example, Raymond wrote that FOSS needed better "ergonomic design and interface psychology, and hackers have historically been poor at these things."[37] Even though KDE and GNOME had been under development for a fair amount of time by that point, user interfaces that could be used by ordinary people, not just hackers, remained a major FOSS weakness.

Microsoft also endorsed a strategy that executives described internally as "embrace, extend, extinguish." First developed to win the browser wars against Netscape (whose investigation by the United States government had brought the tactic to public attention), this strategy entailed integrating community-developed, standards-based technology into Microsoft products, extending those standards with proprietary extensions to attract customers to the Microsoft implementation, and finally leveraging the proprietary extensions as a way to stifle competitors' products and corner the market.[38] Microsoft's adoption of this tactic rather than a campaign based solely on promoting the functionality of its products relative to FOSS alternatives suggested that the company believed that FOSS was sufficiently mature by the late 1990s to compete successfully with proprietary alternatives in areas where it mattered most, despite the usability criticisms noted above.

Proof of Microsoft's plans for thwarting the momentum of the FOSS community via a campaign based both on publicizing usability issues and embracing, extending, and extinguishing open protocols arrived in October 1998, when Raymond obtained a copy of an internal company memorandum. Written by Microsoft product manager Vinod Valloppilli at the request of senior vice president James Allchin for the attention of Paul Maritz, another senior vice president, it was titled "Open Source Software: A (New?) Development Methodology" and dated August 11, 1998. The document revealed the extent and seriousness of Microsoft's concerns over FOSS development and warned executives at the company that "OSS," or open source software, "poses a direct, short-term revenue and platform threat to Microsoft, particularly in server space. Additionally, the intrinsic parallelism and free idea exchange in OSS has benefits that are not replicable with our current licensing model and therefore present a long term [sic] developer mindshare threat."[39]

The paper then detailed several specific ways that FOSS development trumped the proprietary model that Microsoft embraced. It said that the FOSS "release-feedback cycle is potentially an order of magnitude faster than commercial software's" and noted "the ability of the OSS process to collect and harness the collective IQ of thousands of individuals across the Internet."

To respond to the FOSS threat, the paper's author recommended targeting "a process rather than a company." It continued, "Linux can win as long as services/protocols are commodities. ... By extending these protocols and developing new protocols, we can deny OSS projects entry into the market."

Recognizing the usability flaws that some users associated with FOSS, the report also noted that "a key barrier to entry for OSS in many customer environments has been its perceived lack of quality." It suggested that ensuring that Microsoft products would be superior in quality to their FOSS alternatives constituted an additional strategy for stifling the FOSS threat. Yet because the rise of FOSS products in the Internet space, including projects like Apache, had provided "very dramatic evidence in customer's [sic] eyes that commercial quality can be achieved/exceeded by OSS projects," the report's author suggested that challenging FOSS on the basis of quality and usability required a careful approach that would entail dismissing the success of FOSS Internet software as merely "anecdotal."

Raymond, who published the report on his website as the first of a series of leaks he titled *The Halloween Documents* because the first were posted near Halloween, adopted an alarmist stance. In comments on the paper, he warned his followers—who were numerous because Raymond was a leading figure in the FOSS world at the time and had played a prominent role in launching the open source campaign—of Microsoft's "sinister" plan to undercut open standards in order to facilitate "the erosion of consumer choice" and "monopoly lock-in."[40]

On balance, Raymond's reaction to the paper—which he used as an opportunity to promote himself because the Microsoft report made extensive reference to his own writings—did not take into account the limitations of Microsoft's assessment of the FOSS threat. The paper primarily identified FOSS as a challenge for Microsoft in the server market but concluded that "Linux is unlikely to be a threat on the desktop." It also suggested that Netscape's Mozilla effort would gain little traction over the long term. From these perspectives, the paper's tone was less consistent with the sky-is-falling picture that Raymond presented.

Nonetheless, the first Halloween document offered the FOSS community proof that Microsoft was concerned about the software it was developing in some key sectors of the market. That realization was reinforced when Raymond published the second Halloween document, which also was written in August 1998 and leaked to him a few days after he posted the first. Titled "Linux OS Competitive Analysis: The Next Java VM?," the second paper focused on the challenge that GNU/Linux distributions presented to Microsoft's business. Like the first report that Raymond published, the second one was dismissive of the likelihood that desktop Linux-based systems would gain wide adoption. Its author also doubted the ability of Linux programmers to continue "to achieve the big leaps the development team is accustomed to" after the kernel had grown more mature and had no more obvious features to gain. However, the paper identified a number of Linux strengths as well and concluded that an embrace-extend-extinguish approach was necessary for Microsoft to contain the GNU/Linux threat.[41]

A press statement from Microsoft's Netherlands division, which Raymond published as the third Halloween document, confirmed the authenticity of the August 1998 reports.[42] In a more detailed response dated November 5, 1998, the company again acknowledged them as genuine while also downplaying their significance. Microsoft called the leaked documents merely "an engineer's individual assessment of the market at one point in time" rather than "an official statement by Microsoft on the issue of open source software or the Linux model."[43]

The FOSS community, however, perceived a real threat and a need to respond in kind by aggressively discrediting Microsoft products. Stallman wrote that "Microsoft has explicitly targeted our community."[44] Perens described the Halloween documents as evidence "that MS will launch an offensive" against FOSS users and developers.[45] And although Raymond at times cautioned fellow FOSS leaders "that we need to be for software quality, not just against something," the perception that the FOSS community was locked in an existential struggle against Microsoft led by the early 2000s to discussions of "the hacker community's perennial war against Microsoft."[46]

### "Shared Source"

Proof of Microsoft's execution of the embrace-extend-extinguish strategy that Valloppilli had advocated against FOSS seemed to arrive by May 2001, when the company announced an initiative it called Shared Source. According to senior vice president Craig Mundie, Shared Source constituted "a balanced approach that allows us to share source code with customers and partners while maintaining the intellectual property needed to support a strong software business." Under the initiative, Microsoft proposed to share source code with licensed customers, though not with the public as a whole, in order to assist third-party developers. Mundie contrasted Shared Source with GPL-licensed software, which he said "has inherent security risks and can force intellectual property into the public domain." He also made the case that the FOSS economic model was flawed because making money based on distribution or services, rather than software itself, would "not generate the revenue needed for major investments in technology."[47]

Other Microsoft executives adopted this line of argument. In June 2001, Steve Ballmer, who had become the company's CEO the year before, declared that "Linux is a cancer that attaches itself in an intellectual property sense to everything it touches."[48] He also suggested that FOSS was fundamentally at odds with business, claiming that "open source is not available to commercial companies."[49] Such statements presumably reflected a calculated effort to frighten users away from FOSS rather than Ballmer's true beliefs because he was undoubtedly aware that the GPL and other FOSS licenses he criticized did not actually prevent the commercial use of code.

Microsoft's efforts to steer consumers toward its Shared Source program and away from FOSS bore less fruit than the company hoped. A survey that Microsoft presented internally in September 2002, which was leaked to Raymond and became another Halloween document, reported that "most respondents had heard only 'very little' about the initiative."[50] Nonetheless, the effort seemed to the FOSS community to confirm Microsoft's intent to embrace, extend, and extinguish not just a particular FOSS product but the FOSS development model as a whole—by coopting it in favor of a process that provided source code to third parties under particular conditions and yet did not make the code publicly available or freely redistributable.

### Proxy War: The SCO Lawsuits

The next phase in the war between Microsoft and the FOSS community began in March 2003, when the SCO Group filed a billion-dollar suit against IBM alleging that IBM had copied code from the System V release of Unix.[51] The SCO Group, which had a long history in the Unix and Linux business, claimed distribution rights over Unix. (Prior to 2002, the company operated under different names, including Caledra, Inc., under which it had sold a GNU/Linux distribution called Caldera during the 1990s.) The scope of the confrontation expanded during the months that followed, when the SCO Group filed additional suits against companies that used Linux, including Novell, DaimlerChrysler, and AutoZone, on the basis of claims that the kernel illegally incorporated code derived from Unix.

From the perspective of the FOSS community, the SCO Group's lawsuits seemed dubious for multiple reasons. First, in June 2003, the company promised in a letter to its customers that it would not sue them if they used its version of Linux, suggesting that the lawsuits constituted a selfish strategy by the company to gain marketshare through legal maneuvering. A second problem was the SCO Group's refusal to put on public display the code that it claimed Linux developers had stolen from Unix, prompting FOSS advocates to launch a "Show Us the

Code" campaign. Third and most damning in the eyes of FOSS supporters, it emerged from a leaked memo in March 2004, which Raymond obtained and posted as part of the Halloween document collection, that Microsoft had supplied financial backing to the SCO Group. Although there was no conclusive evidence that Microsoft did so with the specific goal of funding the lawsuits related to Linux or that top-level Microsoft executives were aware of the operations, Raymond and other hackers concluded as much, especially after the SCO Group confirmed the authenticity of the document Raymond had posted.[52]

The various legal battles continued for several years until August 2007, when a judge in the case between the SCO Group and Novell ruled that the SCO Group could not claim rightful ownership over the Unix copyright. Although that decision did not resolve the other lawsuits, some of which remain open today, it shattered the foundation of the SCO Group's major claim involving the improper integration of Unix code into Linux. The FOSS community welcomed the ruling as a decisive victory, which proved that "Linux is a safe solution and people can choose it with that in mind," in the words of Jim Zemlin, executive director of the Linux Foundation.[53]

### Samizdat

At the same time that Microsoft was funneling money to the SCO Group, the company also supported the Alexis de Tocqueville Institution, a Washington, D.C.-based research organization that was preparing a book written by Kenneth Brown called *Samizdat: And Other Issues Regarding the "Source" of Open-Source Code*. (*Samizdat* is a Russian word that refers to the clandestine distribution of dissident writings within the

Soviet Union.) Drawing his conclusions largely on the basis of the notion that a kernel like Linux, even in its first iterations, was too complex for one person to have written independently, Brown contended that Torvalds must have plagiarized much of the Linux source code from Tanenbaum's Minix operating system. More generally, his work criticized the FOSS community for peddling what Brown called "hybrid code," by which he meant programs that integrated code from a variety of sources and did not always give proper credit to all of the programmers who had contributed.[54]

Brown's thesis, which was never actually published in book form but appeared in a prerelease report that the Alexis de Tocqueville Institution posted online in 2004, faced rejection by the FOSS community. Tanenbaum himself, whom Brown interviewed in Amsterdam, reported in an essay he posted online after the meeting that "Ken Brown doesn't have a clue what he is talking about." Tanenbaum strongly dismissed the idea that Torvalds had produced Linux by copying Minix code on several grounds but especially because the monolithic Linux kernel differed so starkly in design from the microkernel that Tanenbaum had written for Minix. Although Tanenbaum stated that Torvalds perhaps had not given as much credit as he might have to the people, including Tanenbaum, whose ideas had influenced the general design of Linux, he concluded that Torvalds "did write Linux. I think Brown owes a number of us an apology."[55]

Tanenbaum also suggested that Brown's project might be politically motivated. He reported that when he asked Brown during his interview whether Microsoft was funding the *Samizdat* book, Brown did not deny it. Instead, he insisted ambiguously that the Alexis de Tocqueville Institution had multiple funding sources.[56] Microsoft later confirmed that it funded the organization, although it claimed that it did not direct its support toward any of the Institution's specific projects.[57]

In addition to Tanenbaum, other major figures in the Unix and FOSS worlds, including Dennis Ritchie and Stallman, repudiated the *Samizdat* report.[58] In June 2004, even Microsoft disavowed Brown's work, calling it "an unhelpful distraction from what matters most—providing the best technology for our customers," although the company remained silent regarding what role it might have played in funding or encouraging the report's creation.[59] Ultimately, the *Samizdat* project ended in embarrassment for the Alexis de Tocqueville Institution, which shuttered in 2006.

### THE MICROSOFT WAR IN RETROSPECT

The collapse of the *Samizdat* project and the failure of the SCO Group's legal campaign against FOSS development placed the FOSS community on firmer ground than it had ever before known. It also represented a positive outcome for FOSS that was remarkably different from the one the BSD community had suffered in the early 1990s. Then, disputes over the ownership of the NET 2 code stunted adoption of the first complete, free Unix-like operating system. The BSD community never fully recovered from these troubles despite eventual legal settlements in its favor. In contrast, in the early 2000s, FOSS developers and users emerged relatively unscathed from the drama that beset them during the Microsoft war.

Those divergent outcomes reflected several key differences between FOSS as it existed in the early 1990s and what it became by the 2000s. One difference was that, by the latter period, there was historical precedence for legal disputes to resolve in FOSS users' favor. That fact likely made it easier for companies and individuals to continue using Linux even as the SCO Group sued over the kernel's code.

At the same time, the enormous commercial importance that Linux and other FOSS technologies had assumed by the 2000s meant that a strong coalition of well-funded companies and organizations could resist legal challenges to FOSS. In the early 1990s, the BSD developers had only the University of California system to come to their aid. Although that intervention eventually proved successful, it was not as powerful as the response that FOSS organizations mounted against Microsoft in the 2000s, when, for example, Red Hat countersued the SCO Group in response to the latter's lawsuit against IBM.

Finally, by the 2000s, the FOSS ecosystem had grown so diverse and was exerting its influence through so many different programs and segments of the market that it could weather even deep-pocketed legal attacks against one of its hallmark products, Linux. Snuffing out the FOSS movement was no longer possible. It had become too big to fail.

The end of the SCO Group's legal campaigns and the failure of the *Samizdat* project did not engender complete harmony between the FOSS community and the world of proprietary software. Legal flare-ups with Microsoft continued, especially in the realm of patents, into the late 2000s. Yet the fact that, by the second decade of the new millennium, FOSS companies were burying their swords and actively collaborating with Microsoft signaled fundamental shifts in the trajectory of the FOSS world. So, too, did the diverse range of new niches, from mobile devices to embedded applications, into which FOSS expanded during those years. These changes and their significance for helping to drive the FOSS revolution toward new frontiers are the subject of the next chapter.

# 6 ENDING THE FOSS REVOLUTION?

SHORTLY AFTER SEIZING POWER in 1799, Napoleon Bonaparte declared to the people of France that the French Revolution "has been settled on the principles with which it began; it is over." The statement reflected the Corsican dictator's eagerness to end the ten years of upheaval that had gripped France since the revolution's start.

Yet declaring the end of a revolution and actually ending one are very different things. In 1799, it still was difficult to bring the revolution to a decisive conclusion. In fact, both Napoleon's dictatorship of 1799 to 1815 and the restoration of the Bourbons to the throne (the new French monarch was the brother of the king who was killed by revolutionaries in 1792) failed to halt the revolutionary impulses in the country. France erupted into revolution again in 1830, 1848, and 1871, and those were only the large-scale upheavals. As recently as 1968, the country was subject to widespread revolt inspired in part by the principles of 1789. In this sense, the French Revolution did not end with Napoleon's 1799 pronouncement.

To a large extent, revolution in France proved difficult for successive generations to end because it was impossible for a single regime to embody the revolutionary tradition of 1789 to the entire country's satisfaction. Whenever one faction claimed to have imposed a political order legitimated by the principles of 1789, another faction arose to challenge it. Within the series of recurring revolts that ensued, the meaning and intended aims of the original revolution of 1789 were continually contested and reformulated.

Other than a violent encounter between Torvalds and a penguin in an Australian zoo (which gave rise to the Linux mascot, Tux), the FOSS revolution lacked the bloodshed of the French Revolution.[1] Yet both revolutions followed a similar trajectory in the sense that bringing them to an end has been a messy, subjective affair. Today FOSS reigns supreme and is popular across the entire technology world. It has become the de facto mode of producing, distributing, and using software for hundreds of millions of people. Even Microsoft now embraces FOSS warmly. From this perspective, the goals of the revolutionaries who began promoting FOSS decades ago—when few people had heard of it and the future of computing seemed to be heading in a decidedly different direction—seem to have been fulfilled.

Yet a struggle continues today between different groups seeking to define what FOSS should be, which behaviors properly fulfill the aims of the FOSS movement, and how much more— if anything—needs to be done to make the world truly safe for FOSS (and the hackers who endorse it). This chapter describes the debates and major developments that have shaped the FOSS landscape since the early 2000s—when FOSS established a solid footing outside of hacker circles. It shows that FOSS has become even more successful than many of its proponents fifteen years ago could have imagined by conquering markets like mobile devices and cloud computing. Yet as this chapter also demonstrates, this growth has added to the debates about the nature of FOSS and whether FOSS's successes have betrayed the original aims of the FOSS revolution.

## FOSS TAKES COMMAND

By the early 2000s FOSS's success had become indisputably clear for the first time to many people, even though it remained "hard to find a computer that doesn't run a Microsoft product," as a journalist wrote at the time.[2] Even if FOSS controlled only tiny shares of most markets, its enduring presence within them reflected its success in the eyes of its advocates. One forum poster wrote in 2001 that "the key to success has already been gained by Linux" because "it is used by the people who matter"—by which the commenter meant those involved in "the advancement of computing."[3] A few years later, Steven Weber deemed FOSS successful enough to write an entire book whose titular aim was to explain "the success of open source."[4]

Yet if observers in the early 2000s measured FOSS's success primarily in terms of its persistent presence as a minority part of the technology world, FOSS's success today is defined by a dominance that far surpasses mere survival. Tens of millions of computers—not just traditional desktops and servers but also mobile phones, "Internet of Things" devices, and virtual application images running in the cloud—now depend mostly on FOSS code. Seventy-eight percent of companies report running "part or all" of their operations on FOSS, a number that has nearly doubled in just the past five years.[5] These are changes that researchers such as Weber and FOSS luminaries like Raymond could barely foresee fifteen years ago.

The following sections explore how FOSS has evolved since the early 2000s by examining five key developments of the past fifteen years: FOSS's endorsement by large companies, including Microsoft, that previously espoused no interest in FOSS or actively combated it; the emergence of the Android mobile operating system; the introduction of Ubuntu GNU/Linux; the advent of the OpenStack operating system for cloud computing; and the use of FOSS in embedded computing devices.

### FOSS and Business

IBM went out on a limb when it announced its support for Linux in the late 1990s. At the time, no other company of comparable size was willing to stake part of its image on FOSS code.

Starting in the mid-2000s, however, moves like IBM's became common for large technology companies. Dell began selling computers with GNU/Linux preinstalled in 2007.[6] In 2009, Google introduced Chromium OS, an open source project to build a lightweight Linux-based operating system for the Web-centric devices that Google sells.[7] HP's CTO in 2015 declared that "open source software" is "part of the fabric of everything we do," referring to the company's operations.[8] These are just a few of the numerous examples of the large-scale endorsement of FOSS in recent years by companies that have not traditionally engaged in the FOSS space.

Of equal historical significance is the friendliness Microsoft has shown in recent years toward the FOSS community. FOSS companies such as Canonical, whose history is detailed below, readily collaborate with Microsoft on cloud computing, technology for the Internet of Things, and more—a move made easier when Canonical's founder, Mark Shuttleworth, declared in 2013 that unseating Microsoft from its leading position in the PC software market was no longer a primary goal of Ubuntu GNU/Linux developers.[9] In fall 2014, Microsoft's new CEO, Satya Nadella, declared that his company "loves Linux."[10] A year later, Microsoft announced that it was building its own Linux-based operating system to help run its internal network of cloud servers.[11]

These kinds of changes did not alter Microsoft's image in the hearts of all FOSS advocates. Some expressed worries that the company might have a "hidden agenda" in launching its own Linux-based operating system, even though the platform was designed only for internal use.[12] Another FOSS advocate wrote, "Despite Microsoft's continued assault on Linux and on Android (using software patents, which it still discreetly lobbies for), some figures in the media are perpetually peddling the Microsoft-serving lie that 'Microsoft loves Linux.'"[13] Yet Microsoft's engagement with the FOSS community remains ongoing. Most recently, the company introduced Linux Subsystem for Windows, a platform that provides binary compatibility between GNU/Linux applications and Windows operating systems—essentially the reverse of what Wine does for Windows applications running on GNU/Linux.[14]

Microsoft and other major technology companies have not endorsed FOSS wholesale; they tend to use FOSS products only for certain purposes. In most cases, they combine FOSS with closed source software in a way that provides end users, such as those using Google Chromebooks or accessing websites hosted on Microsoft's Azure cloud, with little inkling that they are relying in part on FOSS software. In this sense, the developments discussed above are little different from the role that Linux and Apache software assumed in the mid-1990s on Web servers, where end users usually knew little about the software hosting the websites they visited. This trend has also fueled concerns that the widespread endorsement of FOSS in the business world is merely an expedient pursued by companies wishing to acquire "both software and the associated research and development ... at very little expense ... without offering anything much in return themselves" to the FOSS community, as Hall has noted.[15]

In addition, most of business activity related to FOSS has focused on the "open source" strand of the ecosystem. Stallman and his free software cohorts generally have remained on the sidelines as FOSS commercial activity reached new heights in recent years. That detail matters little to the organizations that deploy FOSS, which they eagerly endorse regardless of whether the Free Software Foundation approves of the licensing terms behind particular products. But it nonetheless highlights how FOSS's widespread adoption by major companies has deepened fractures within the FOSS community, making it harder for FOSS hackers to reach consensus about whether their revolution has achieved its goals.

### Android

Another key marker of FOSS's rise to dominance within certain markets in recent years is Android. Android is the mobile operating system that has placed the Linux kernel and some other key FOSS programs (such as the WebKit rendering engine) in hundreds of millions of smartphones and tablets in recent years.

In the broadest sense, Android's history dates to 2003, when a team of California entrepreneurs launched Android, Inc. Their initial goal was to develop software for digital cameras.[16] In 2005, Google acquired the company, which at the time remained an obscure venture.[17] Google put Android, Inc. developers to work building an operating system for phones that was based on the Linux kernel and adaptations of some other FOSS utilities. At the same time, Google courted industry partners to help launch what it saw as an important competitor to closed smartphone operating systems, especially Apple's iOS for iPhones.

Google's programming and partnership efforts came to fruition on November 5, 2007, when it announced Android to the world. The company billed the operating system as "the first truly open and comprehensive platform for mobile devices."[18] At the same time, Google launched a partner network called the Open Handset Alliance that it built to promote and distribute Android. The next year, T-Mobile introduced the first Android-based smartphone, the G1. Other carriers quickly followed suit. Google's own Android-powered phone arrived in 2010, when the company began selling Nexus smartphones online. Soon Google's Linux-based operating system became a major contender in the mobile operating system market, in which Android powered 80 percent of devices by 2013.[19] In 2016 the figure reached ninety percent.[20]

That was the type of market-share conquest that FOSS developers in other niches could only dream of. On Web servers—which represent the second-greatest FOSS success story—the market share of Apache software running in conjunction with Linux peaked at around 70 percent in the mid-2000s and has since declined to below 40 percent today (some of that market share has been replaced by NGINX, another open source Web server).[21] FOSS software's penetration of the PC market, a longstanding focus of many FOSS advocates, has been much smaller. Realistic estimates suggest that no more than about 5 percent of desktop computers worldwide have ever run GNU/Linux platforms.[22] Against these figures, Android's sustained conquest of more than four-fifths of the global market for mobile operating systems is spectacular.

Yet not all FOSS supporters welcomed this feat. Because Google and the Open Handset Alliance placed little emphasis on the FOSS core of the Android platform, some FOSS advocates viewed Android as a threat to openness and interoperability more than an outsize example of FOSS's success.

The decision by Google and its partners to downplay the importance of FOSS to Android was no mistake. From the beginning, Google was circumspect about advertising the platform as a distribution of Linux. The company did not mention the kernel in the official announcement of the platform in 2007. By 2009, Google engineer Patrick Brady publicly declared that "Android is not Linux," primarily because it does not include the standard "Linux utilities" (most of which were actually GNU programs, not the work of Linux kernel developers) and lacks support for the glibc C library.[23] Instead, Android uses a customized C library called *bionic*, which is derived mostly from the BSD code base.[24]

For the most part, Google executives even avoided using the term *open source* in describing Android, despite their extensive emphasis on the "openness" of the platform. To them, *open* referred to the collaborative nature of the development and marketing ecosystem surrounding Android rather than the code itself. Most other organizations that belonged to the Open Handset Alliance similarly stuck to the *open* term without referencing Linux or open source software when Android was announced. Texas Instruments and Wind River were the only exceptions.[25]

For licensing, Google and its partners adopted permissive Apache-style terms for most of the Android platform. The Linux-based kernel code, which they could not legally switch from its original GPL license, was the only major part of the Android code base that remained subject to licensing terms that aggressively protected the openness of the code.[26] Thus, just as some Linux proponents had failed to acknowledge the importance of GNU code to GNU/Linux distributions in the 1990s, Google and most of its partners gave short shrift to the hacker community and culture that had unwittingly done so much to make Android a reality.

Both politically and technically, the disconnect between Android and the mainstream FOSS community irked many hackers. After the November 2007 Android announcement, critics contended that the Apache-style licensing of most of the Android platform stunted the potential of Android to become a truly innovative operating system for mobile devices by encouraging more FOSS development in that market. One critic wrote that

> Google has sacrificed an opportunity to encourage greater openness in the broader mobile software space. If Android was distributed under [version 2 of the GPL], companies building on top of the platform would have to share their enhancements, which

could theoretically lead to widespread sharing of code and a more rapid acceleration of mobile software development.[27]

Others complained that "the use of the Apache license is the biggest problem with Android."[28] Such remarks reflected worries that the permissive licensing of most of the Android code meant that "manufacturers might fork the code road in a non-interoperable kind of way" by building their own Android variants without sharing their modified code with the Open Handset Alliance members or the FOSS community more broadly.[29]

Google responded to such criticisms by requiring members of the Open Handset Alliance to agree not to "fragment" the code by releasing mutually incompatible variants of the platform.[30] That mandate, however, did nothing to prevent companies that did not belong to the group from doing as they wished with Android. Nor did it placate users who argued that "if Android had just used the GPL (which prohibits forking), then this problem would have [been] avoided."[31]

This argument was flawed; nothing in the GPL prohibits software forks. Nonetheless, the idea that Android's openness would have been assured if only Google had licensed all of the code using the GPL reflected the persistent belief among some hackers that the GPL alone, rather than alternative FOSS licenses, could facilitate the goals of the FOSS revolution.

Leading hackers have also been critical of the way Android incorporates FOSS code. Although Stallman in September 2011 called Android "a major step towards an ethical, user-controlled, free software portable phone," he lamented that Google had originally refused to release the source code of the non-GPL-licensed components of the platform. He also complained that Android makes it easy for hardware manufacturers and software programmers to incorporate nonfree software applications into Android-based devices.[32]

Linux kernel developers excised the Android driver code from the kernel in 2009, a decision they said they were "so sad" to have to make.[33] That change effectively divorced Android from the mainstream Linux kernel code base, a situation exacerbated by Google programmers' decision to develop their own solutions for tasks such as power management on Android rather than borrowing the code from Linux.[34] By March 2011, in another example of confusion over the GPL's provisions regarding software forks, some critics speculated that Google might face legal action for having violated the GPL by forking the Android code base.[35] Torvalds dismissed such claims as "totally bogus," and he has since expressed optimism that the Android and Linux code bases will eventually return to a state of mutual compatibility.[36] Still, it was clear since Android's introduction in 2007 that many FOSS hackers were unhappy with Google's approach to Android and the practical lack of opportunities for collaboration that the forking of the Android kernel code has imposed.

### Ubuntu

Android is not the only Linux-based operating system whose impressive success within the marketplace has outpaced the enthusiasm it generated among hackers over the past decade. Ubuntu, which emerged as one of the most popular GNU/Linux distributions of the 2000s and 2010s, followed a similar trajectory.

From a technical perspective, nothing sets Ubuntu apart in a fundamental way from other GNU/Linux distributions. Launched in 2004, it is much newer than other distributions of enduring commercial importance, such as Red Hat and SUSE. But especially in its early years, the Ubuntu code base was little more than a spinoff of Debian GNU/Linux. (In the 2010s, Ubuntu acquired a distinctive technical profile by adopting original components like the Unity desktop environment.)

In nontechnical ways, however, Ubuntu bears little resemblance to other GNU/Linux distributions. One important distinction is the unique circumstances surrounding its founding. In contrast to most of the hackers who launched the other major GNU/Linux distributions, the South African who created Ubuntu, Mark Shuttleworth, had a background in finance and cybersecurity, not operating-system design or Unix. While GNU/Linux distributions like Red Hat and Debian were appearing in the mid-1990s, Shuttleworth was founding a company called Thawte. Launched in 1995, its main business was to provide digital certificates that websites require to encrypt online content. Shuttleworth became a millionaire four years after starting the company by selling it to VeriSign for the equivalent of about $575 million.

Shuttleworth spent part of his fortune paying for a trip into space, which he completed in 2002. By 2003, however, when he began writing on his blog about funding free software development, he had become interested in FOSS.[37] The next year, he founded Canonical as a private for-profit company to sponsor development of a new GNU/Linux distribution. That system became known as Ubuntu, a name that derived from words in the Xhosa and Zulu languages that signify "solidarity" and "allegiance."[38]

Shuttleworth's attraction to FOSS was remarkable because there was little in his background to suggest that he would become a major figure in the FOSS community. Born and raised in South Africa, he was far removed from any of the geographic centers of FOSS development. As he tells the story, he was "not at all" a follower of the Free Software Foundation or Linux development before the period when he became interested in the Ubuntu project. In fact, his first experience with Linux happened when he "snuck into the University of Cape Town," where he was a student, "late one night with a key I wasn't supposed to have" and installed GNU/Linux—from a pile of floppy disks—on a lab computer in place of Windows. (He reinstalled Windows before morning to cover his tracks.) Yet despite this early experimentation with FOSS, Shuttleworth's company had little involvement with Linux or other FOSS projects, although it did use some Apache-licensed software.

What eventually caused Shuttleworth to focus on FOSS after the success of Thawte, in his own telling, was his belief that GNU/Linux and similar software could provide a foundation for helping to improve society. He told me that by 2004, "I didn't have to work, I wanted to help people do great things" and FOSS seemed the best vehicle for achieving that goal.[39]

Toward that end, Canonical unveiled the first version of Ubuntu, 4.10, in October 2004, inaugurating a tradition of assigning Ubuntu version numbers according to the month and year of each release. The release was also known as "Warty Warthog" because its lack of refinement made it "wartier" than Shuttleworth hoped Ubuntu would eventually become. Accompanying Ubuntu 4.10 were a series of philanthropically informed promises by Canonical. The company declared that Ubuntu would always be available free of cost; that the system would deliver "the very best in translations and accessibility infrastructure that the Free Software community has to offer, to make Ubuntu usable by as many people as possible"; that new versions of Ubuntu would be released according to a regular and dependable schedule; and that the Ubuntu team was "entirely committed to the principles of open source software development; we encourage people to use open source software, improve it and pass it on."[40]

In sum, the Ubuntu promise was to empower people by providing them access to software they might not otherwise have and by promoting the philosophy of the open source world. (Although Shuttleworth in 2003 used the term *free software*, by 2004, the terminology within the Ubuntu camp had shifted primarily to *open source*.) Through the Ubuntu initiative, Shuttleworth hoped to resolve what he called "a deeply unhealthy situation in personal computing," which was Microsoft's near-monopoly in the early 2000s over the operating-system market for PCs.[41] As a result of that situation, Shuttleworth believed, access to affordable quality software was limited, as was the availability of programs that catered to users outside of dominant linguistic or cultural demographics.

To see the Ubuntu mission through to completion, Shuttleworth and Canonical adopted a pragmatic strategy. Although Shuttleworth assigns a great deal of credit to Stallman and the GNU developers for introducing much free code to the world, he told me they "lost touch" long ago with the momentum and purpose of free software development. He said that if most FOSS developers adhered only to Stallman's principles, the FOSS community "would be stuck in a world of ideological pain" and unable to produce viable software.[42]

Shuttleworth was certainly not the first person to criticize the free software camp for an ostensible lack of pragmatism. That debate had raged throughout the 1990s, as the previous chapters show. Yet what made Ubuntu and Canonical different from the FOSS initiatives that preceded them was just how far Shuttleworth and his collaborators were willing to go in prioritizing pragmatism over free software ideology. Although the Ubuntu distribution has always been available under open source licenses, Canonical introduced a number of related tools, such as the Landscape server-management service and parts of its Ubuntu One file-syncing service, as proprietary software. (It eventually made some of these products open source, but not until they were already in widespread use.) The company also expressed no qualms about integrating bits of closed source firmware—known to hackers as *binary blobs*—into Ubuntu in order to power hardware devices that would not function without it. And in recent years, Canonical has opted to internalize much of the Ubuntu development effort by requiring programmers to work behind closed doors rather than in the public FOSS bazaar, even for code that it intends eventually to release as FOSS.[43]

By many measures, Ubuntu developers' ultrapragmatist take on FOSS has served the Ubuntu community well. Canonical has succeeded in issuing timely new versions of Ubuntu every six months for more than a decade without fail, excepting Ubuntu 6.06, which appeared in June 2006, two months later than initially planned. Few other FOSS projects can match that record of punctuality, which no doubt owes much to the decision by Shuttleworth and his collaborators to ignore political distractions related to free software ideology.

Ubuntu has also proven popular among users. Canonical's status as a private company means its financial records are not available for scrutiny, and reliable statistics on the number of people using Ubuntu, which is distributed for free and requires no registration, remain elusive. Nonetheless, the available evidence suggests that Ubuntu has seen remarkable levels of adoption in the thirteen years since its first release. Canonical estimated in 2015 that Ubuntu's user base totaled 40 million people and reported (presumably on the basis of server statistics to which the company has access) that thirty thousand Windows users download Ubuntu each day.[44] Governments and other organizations, especially in Europe, have installed hundreds of thousands of Ubuntu systems.[45] On servers, Ubuntu was the second-most popular GNU/Linux distribution for hosting websites in 2015, and it runs more than half of the virtual servers in Amazon's popular EC2 cloud computing platform.[46]

Much of Ubuntu's success stems from the unique advantages it enjoys over competing GNU/Linux distributions that lack supporters who are as wealthy as Shuttleworth, who remains centrally involved in Ubuntu design today despite having stepped down as chief executive officer of Canonical in 2009. Thanks to Shuttleworth's fortune, Ubuntu benefits from a purpose trust, the Ubuntu Foundation, which Shuttleworth founded in 2005 and seeded with $10 million. The trust ensures that the Ubuntu project will have a financial cushion that is independent of Canonical.[47]

Ubuntu adoption also received a boost from a program called ShipIt. From 2005 until 2011, when Canonical discontinued the initiative, ShipIt made Ubuntu CDs available through the mail at no cost worldwide. The program helped to ease the adoption of Ubuntu for users who otherwise would have lacked the Internet bandwidth or technical expertise required to download an Ubuntu installation CD on their own or the cash to purchase one from another distributor.[48]

In Shuttleworth's own estimation, Ubuntu's focus on usability also does much to explain the distribution's impressive growth. "The fact that Ubuntu just works is enormously valuable to millions of people," he said in an interview. He believes that this characteristic sets Ubuntu apart from other GNU/Linux distributions, which have sometimes privileged factors other than usability.[49] Shuttleworth also attributes the distribution's success to the strong commercial backing it received from its association with Canonical, a relationship that enables Ubuntu to "do things Debian can't, and vice versa" because distributions like Debian have no corporate backing. Lastly, he contends that the ability of the Ubuntu leadership, including himself, to be blunt and make dictatorial decisions when necessary has kept the operating system in good health.[50] "I think that one of the reasons Ubuntu has kept moving," despite having grown into a large and complex project, is that its leadership has been willing to "put people off" if necessary, he said. In contrast, "too many good projects" in the FOSS space "have been completely hijacked because they didn't have any mechanism to calm" tensions between competing factions within the generally decentralized infrastructure of FOSS communities.[51]

Many FOSS users have praised the perceived usability and predictability of Ubuntu. "This product just seems to work," one user wrote of Ubuntu in response to criticism of Canonical for having made few code contributions to the Linux kernel. He continued, "So, to that end I do thank Mark Shuttleworth for his efforts and I hope he realizes that he has made others' lives better."[52] Another called Shuttleworth "some kind of Heinlein-esque hero," referring to the science fiction author, Robert Heinlein, who popularized the notion of "paying forward" one's success by helping others to achieve the same positive outcomes as oneself.[53] Torvalds has praised Ubuntu for its "very user-centric" approach to computing.[54]

Ubuntu does have its critics, however. Stallman has complained that when Ubuntu developers "present convenience as their goal rather than freedom, they're teaching people not to value freedom." He also has described the operating system as a form of "spyware" because of features that communicate data about users' behavior with Canonical for marketing purposes, a concern shared by users who complain about Ubuntu's "spyware-inspired keylogger."[55] Online commenters have wondered whether Ubuntu chose to pursue a policy of "selling out" by focusing on profit rather than on Canonical's ostensible mission of making computers more accessible.[56]

Similarly, Canonical's attempt in 2009 to introduce an "Ubuntu Store" feature into the operating system, through which users could download both free and for-purchase programs, prompted a revolt within the Ubuntu community. One user complained that the effort "inclines me towards thinking [Ubuntu is] a for-profit venture, and I guess that's because the word 'store' is now tainted by all the proprietary software repos[itories] out there that have previously been listed."[57] Another, confused by news that Canonical did not actually plan to sell software in the Ubuntu Store at the time but instead aimed to offer only free downloads, asked, "Doesn't 'store' imply that the user will be paying for the software?"[58] Canonical eventually changed the name of the feature to "Ubuntu Software Center," reflecting the company's awareness that not all of its supporters were comfortable with what some viewed as stark departures from the ideals of free software.

Ultimately, Ubuntu's legacy is undeniable as a GNU/Linux distribution that introduced FOSS technology to millions of people who otherwise may never have used it in a prominent way. Yet its effectiveness in advancing the FOSS cause has proven more controversial. Through Ubuntu, Shuttleworth fulfilled his goal of providing "Linux for human beings" (the original tag line of the distribution), and Canonical's focus today on "converging" Ubuntu across multiple types of hardware profiles—including PCs, servers, phones, and televisions—extends that mission by aiming to provide a common, FOSS-based user experience on all of the digital devices on which users now rely.[59] But in the eyes of hackers such as Stallman, Shuttleworth has accomplished little to bring the FOSS revolution to completion.

### FOSS and Cloud Computing

Similar tensions exist within the world of cloud computing, the most significant new frontier that FOSS has entered in recent years. The term *cloud computing* refers to managing data by using software on servers hosted on the Internet rather than running the software directly on local devices. In a sense, cloud computing has existed since computers were first networked together in the late 1960s to allow people at one computer to access computational and data resources from another computer remotely. The World Wide Web, which lets users connect to remote servers in order to retrieve or process information, has also essentially been a cloud-based service since its earliest years.

Yet not until the mid-2000s did organizations embrace the cloud as a mechanism for delivering sophisticated types of software programs. Over the last decade, for instance, it has become common to edit documents via the cloud using Web-based word processors, a type of program that traditionally worked only on local computers. Companies also now commonly leverage the cloud paradigm as a way to expand their computing infrastructure without having to buy and maintain equipment on their own premises. They create cloud-based servers that run as virtual machines and can be set up almost instantaneously. Management of the virtual machines can be outsourced to cloud service providers, reducing staffing costs for the company itself. In this and other ways, the cloud has become a vital part of the computing experience for businesses as well as end users who rely on computers to store information, complete work, watch videos, or perform other common tasks.

In two key ways, the cloud is home territory for FOSS. First, the cloud entered into widespread use in the 2000s, after FOSS had already proved its mettle and programs like the Apache Web server and Samba had demonstrated how effectively FOSS code could help to deliver information over the Internet. Second, because the cloud blends together servers, storage software, and desktop computers that often run different types of operating systems, it demands a high degree of interoperability. Open standards and source code that can be freely shared are therefore a vital resource for building clouds that are agnostic with regard to the types of devices that comprise or connect to them.

Executives at Rackspace, a company that sells access to virtual servers running in the cloud, had interconnectivity in mind when they announced a major Apache-licensed platform in July 2010 "to help drive industry standards, prevent vendor lock-in and generally increase the velocity of innovation in cloud technologies."[60] Called OpenStack, the project was based on code created by Rackspace for its hosting business and technology from Nebula, a hosting platform used by the National Aeronautics and Space Administration (NASA). For its first release in October 2010,[61] OpenStack had two major components—Object Storage, which allowed users to create storage servers designed to communicate data over the Internet, and Compute, which managed computational resources for virtual cloud-based servers. Together, these components made it possible to build clusters of virtual servers running in the cloud, which organizations could use for tasks like storing information and running Web-based applications.

OpenStack has grown steadily since its introduction. Today it includes several components not available in 2010, such as programs for managing networking resources on clusters of cloud servers. The number of organizations and individuals developing OpenStack code has also risen tremendously since Rackspace launched the project in 2010. In April 2016, more than 110 companies and 1,100 individuals were active OpenStack contributors.[62] Although analysts continue to debate whether OpenStack is yet ready for large-scale enterprise deployment, the platform is indisputably the most important FOSS component of the evolving cloud-computing ecosystem.[63] Combined with the importance of open standards generally within the cloud, OpenStack seems to have ensured that FOSS will play a prominent role in this increasingly important market and eliminated the risk that proprietary code will end up powering the cloud.

Nonetheless, some FOSS leaders remain wary of the cloud in general. Chief among them is Stallman, who warned in 2008 that the cloud computing paradigm is "worse than stupidity."[64] Part of his criticism centered on the perception that the cloud was more of a marketing fad than a meaningful reality. This was a fair argument in light of the fact that, as noted above, cloud computing was not actually a new idea when usage of the term exploded in the mid-2000s.

Yet Stallman also expressed deeper anxieties about the cloud. In certain situations—particularly those that he described as "Service as a Software Substitute," mocking the popular "Software as a Service" term that companies use to market some cloud products—software programs running in the cloud "wrest control from the users even more inexorably than proprietary software," Stallman wrote.[65] He meant that because users of cloud-based programs do not have control over the server on which the programs run, cloud users place their data and computing resources at the disposal of whoever owns the cloud server to which they are connecting. Even cloud-based applications that are composed of FOSS code force users to surrender personal data to someone else's server, Stallman has noted.[66]

Complicating matters is the fact that cloud-based software code cannot be regulated in the same way as software that is designed for end users to run directly on the computers they use. The GPL contains a provision that allows developers to use and modify GPL-licensed code without sharing their modifications with the public if they do not distribute the software publicly. Because the cloud makes it possible to run a program on a private server while allowing third parties to access that program over the Internet, it gives rise to ambiguity in determining whether such a program has been publicly distributed and is therefore subject to the GPL mandate that publicly shared software must come with source code. To address this issue, the Free Software Foundation in 2007 introduced a special license, the GNU Affero General Public License, which was based on a license created in 2002 by a FOSS funding organization named Affero.[67] The license provided a way for FOSS supporters to help prevent abuse of the GPL in the context of the cloud, but to date it has seen relatively limited adoption. Moreover, as the Free Software Foundation emphasizes, the Affero GPL does not protect users against the loss of control over their computing environment that occurs when they access software programs through the cloud. It only helps programmers to keep their code open.

For these reasons, the prominent role played by FOSS projects like OpenStack in the cloud computing market has not made the cloud a FOSS-friendly environment in the eyes of figures such as Stallman. Like Android and Ubuntu, cloud computing in some ways has undercut the goals of the FOSS revolution and created challenges that FOSS enthusiasts could not envision when they began creating free software decades ago.

### Embedded Computing

A similar trend has occurred in the embedded-devices market. This sector involves special-purpose hardware whose functionality is generally limited to a small set of specific operations. The devices usually lack the resources to run traditional operating systems. Like the cloud, embedded computing is by no means a new idea; minimalist hardware devices have long existed. However, the increased technological sophistication of devices (such as home appliances) and the demand for the ability to control hardware remotely (via the Internet of Things) have made embedded computing more visible in homes and offices in recent years. Because the openness of FOSS code makes it easy for device manufacturers to keep their code lean (by selecting only the software components needed to work with a certain type of device) and to modify the components if necessary (to ensure compatibility with their hardware), all without paying licensing fees, FOSS is ideal for vendors in the embedded-device market.

Embedded devices have created new opportunities for FOSS software like Linux, which now commonly powers such hardware as wireless routers, "smart" TVs, and TomTom GPS products. Meanwhile, consortiums of researchers, nonprofit groups, and businesses are currently investing millions of dollars to build FOSS solutions for embedded-computing applications in hardware as diverse as drones and automobiles.[68]

Yet if embedded computing has provided new opportunities for FOSS to grow, the limited functionality of the hardware has meant that users often have little control over their computing experiences, no matter how free and open the code that powers their devices is. A smart thermostat that runs Linux is unlikely to provide an interface that would allow an ordinary user to modify the way the device works or to customize its software. The FOSS code that powers a portable music player cannot be easily copied or shared, even if software licenses make it legally permissible to do so.

To make matters worse, embedded devices frequently pair FOSS programs with proprietary software, especially hardware-specific device drivers, whose code is not open. This trend does not sit well with developers interested in an open software ecosystem. As security researcher Bruce Schneir wrote about Internet of Things devices in 2014, for instance, "many of the device drivers and other components are just 'binary blobs'—no source code at all. ... We need open-source driver software—no more binary blobs!"[69] Similar challenges exist on traditional computers, where certain hardware components, like wireless cards, have sometimes required proprietary drivers to operate properly even if the rest of the software on the system is open. In these cases, dedicated FOSS programmers have generally found ways to write reverse-engineered FOSS code to make the devices work, although this can raise legal issues.[70] However, the great diversity of the types of processors, storage hardware, and other device components in the embedded-computing sector makes reverse engineering a less feasible solution there.

Because embedded computing favors the use of FOSS but limits the ability of users to inspect or modify the software that powers them, it also constitutes both a challenge and a boon for the FOSS revolution.

## BEYOND SOFTWARE: FOSS AND FREE CULTURE

A different story has emerged within the realm of the "free culture" movement, where the effects of the FOSS ethos on society as a whole have become clearer than ever in recent years. Here, perhaps more than in the realm of software itself, the

FOSS revolution's legacy will endure most prominently over the long term.

Broadly defined, the term *free culture* refers to the idea that, to maximize the creative and productive potential of individuals, society should remain open, transparent, collaborative, and unfettered by the restrictions of traditional copyright or arbitrary hierarchy. The idea is not a new one. As one of the free culture movement's chief proponents, Lawrence Lessig, noted in his 2004 book on the topic, "the norm of free culture has, until quite recently, and except within totalitarian nations, been broadly exploited and quite universal."[71]

Yet since the turn of the new millennium, the Internet and other new forms of media, along with attempts to control them or rein them in, have vaulted the debate over free culture into the public sphere. They also have prompted the creation of organizations to promote free culture that take their cues directly from the FOSS movement. These groups apply FOSS principles to issues far outside the realm of software.

The most prominent of these organizations is Creative Commons. Lessig, a law professor and former board member of the Free Software Foundation, founded Creative Commons in 2001 with a small team of collaborators. He was inspired by Stallman, whom he credited with having first developed "all of the theoretical insights" that Lessig later associated with the free culture movement.[72] That attribution was unsurprising. As chapter 3 notes, beginning in the 1980s, Stallman and the GNU project endorsed and supported such endeavors as Project Gutenberg and the Open Book Initiative even though those organizations had nothing directly to do with software development or distribution. The relationship between the FOSS movement and the free culture movement dates to FOSS's earliest days.

Yet Creative Commons was novel in that it was the first organization to apply, in a systematic way, ideas that had been born in the FOSS community to society more broadly. To do this, the group developed a series of intellectual property licenses, modeled on the GPL, that writers, artists, and other producers of creative materials including but not limited to code could use to "copyleft" their work. Like the GPL, the licenses, which Creative Commons began releasing in 2002, were designed to provide "an alternative to traditional copyrights by establishing a useful middle ground between full copyright control and the unprotected public domain," according to one early analysis.[73]

Over the past fifteen years, Creative Commons' activities have expanded to include efforts to build more open and collaborative models for scientific research, promote the sharing and reuse of educational materials, and facilitate open collaboration in the realms of culture, government, and science. The organization has endeavored to do for public life as a whole what FOSS hackers did for software—challenge traditional, closed-access modes of production and supplant them with ones in which all users are free to participate on the basis of their demonstrated merits, unencumbered by bureaucratic or hierarchical barriers that privilege the interests of particular groups over those of the public in general.

*Wikipedia*—the free online encyclopedia that volunteers have collaboratively developed via the Internet since its launch by Jimmy Wales and Larry Sanger in 2001 (shortly after Stallman published an essay calling for a free, online encyclopedia)—is another prominent example of a project that has extended FOSS principles into new territory.[74] By allowing anyone to write and edit articles and relying on community consensus to determine what material becomes available for public consumption, *Wikipedia* adopts a production strategy for text that is similar to the bazaar-style development model behind most FOSS code today. The encyclopedia, which the historian Roy Rosenzweig compares to Linux and other FOSS projects, shares a direct affinity with both GNU and Creative Commons by licensing most of its articles under the GNU Free Documentation License and the Creative Commons Attribution-ShareAlike License.[75]

Economically, FOSS has exemplified and helped to advance new modes of production with positively revolutionary consequences for society. As Yochai Benkler has persuasively argued, the new "networked information economy" has placed "the most important components of the core economic activities … in the hands of the population at large" for the first time since the Industrial Revolution.[76] FOSS projects are only one way that distributed modes of production, powered by computer networks, have challenged traditional economic models that centralize wealth and the means of production at the top of an industrial hierarchy. The advent of what Benkler calls "open source economics" cannot be attributed to FOSS alone. Yet FOSS projects are a prime example of this phenomenon, and the code that FOSS developers produce is deeply intertwined with the technology that makes the new information economy function. In these regards, the networked information economy is another area where the effects of the FOSS revolution are playing out beyond the realm of software.

## FOSS AND DEMOGRAPHIC DIVERSITY

Despite the wide-ranging effects that FOSS has exerted on economic and cultural practices within society as a whole, one puzzling trend has persisted within the FOSS community. This is the lack of diversity among FOSS developers and, to a lesser extent, users.

In theory, FOSS development models make it much easier than the alternatives for anyone to contribute to software production, regardless of race, gender, nationality, age, professional status, or any other outward characteristic. As chapter 1 notes, the principle that "hackers should be judged by their hacking, not bogus criteria" (like innate traits) is baked into the hacker ethic itself, at least as Levy articulated it in 1984.[77] And the fact that FOSS programmers most often collaborate not in person but via electronic channels, where participants often cannot infer the races or genders of their counterparts, suggests that FOSS should be an especially color- and gender-blind space.

With few exceptions (most notably Mitchell Baker, a woman who has been a key figure in Mozilla since the late 1990s), however, FOSS has tended to remain the realm largely of white men. These men are diverse in terms of their national origins, but otherwise, their dominance of the FOSS space appears to come at the expense of women and minority contributors who might otherwise play a greater role.

This trend may seem unsurprising given that female and minority professionals have long faced underrepresentation in the realms of software and technology in general. United States government statistics for 2015 show that blacks account for only 5 percent of software developers, compared to 11.7 percent of the total workforce. Women comprise 46.8 percent of the workforce but only 17.9 percent within the software industry.[78] Yet the FOSS world is remarkable in that it "is even whiter and more male than the world of proprietary software," as one of the few journalists to examine this phenomenon notes in a 2013 article.[79]

The FOSS community has long quietly acknowledged this trend. For example, in response to a headline that proclaimed "Open Source Geeks Considered Modern Heroes," a Slashdotter in 2004 cautioned, "But be sure to read the small print: Exception: women."[80] Very little research has attempted to explore the reasons behind this reality. One factor may be that minorities tend to have lower levels of education and less access to the Internet than their white counterparts, making it harder for them to enter the FOSS space, centered as it is on academia and the Internet.[81] At times, leading FOSS figures have been less than welcoming toward women and minority groups. For example, in one of his more obscure writings, Raymond compared female computer scientists to "amazons," "bimbos," and "impossible anonymous synthetic blondes in an upscale skin magazine," imagery he apparently found "much less threatening" to embrace than the reality that there could be highly competent, professional women programmers.[82] More recently, Torvalds stated publicly that "the most important part of open source is that people are allowed to do what they are good at" and "all that [diversity] stuff is just details and not really important."[83]

More significant than the negative messages that arise from statements such as these, however, are the economic realities of the FOSS world. Especially in the early days of the FOSS revolution, few FOSS developers were paid. Instead, in the vision of leading FOSS theorists like Raymond, they operated in a "gift culture" in which compensation for the time and effort they put into writing FOSS code came in the form of reputation.[84] The scholarly analyses of FOSS communities by Weber and Kelty generally support this interpretation.[85]

For developers economically privileged enough to be able to prioritize the esteem of their peers over material rewards, working within the FOSS gift economy is not an unreasonable proposition. For minority and female developers who on average are less well paid and possess fewer financial assets than white males, however, the suggestion by writers like Raymond that FOSS contributors should be happy to work only in exchange for reputation could be difficult to swallow. Lacking sufficient financial resources or income from other means to be able to write FOSS code for free, many minority and female programmers might have found it difficult to enter the field or to sustain the level of coding contributions necessary to establish a reputation among fellow FOSS hackers.

This argument, however—that female and minority programmers face a harder time entering the FOSS world because it offers less financial compensation—is increasingly difficult to accept. In recent years, the demand for professionals with skills in technologies such as Linux has surged, and the pay offered to them has risen at a rate nearly double that for the technology sector as a whole.[86] FOSS developers no longer have to be willing to accept little or no monetary compensation for their work. Moreover, even though a lack of payment for FOSS work has historically discouraged women and minority programmers from contributing, the low cost of FOSS software compared to

most proprietary alternatives should have persuaded less afflu-
ent users to adopt it. For these reasons, economics alone does
not explain the lack of demographic diversity within the FOSS
community today.

Understanding the issue fully requires paying close attention to certain philosophical dimensions of FOSS development.
Tara McPherson has argued that Unix itself—although designed
by programmers who generally sympathized with the progres-
sive political reforms that accompanied Unix's evolution in the
1960s and 1970s—embeds notions of segregation within its
core design principles. In McPherson's view, the modular design
of Unix—the idea that an operating system should be composed
of independent parts, each one designed to do a specific job and
do it well—reflects "the way in which the emerging logics of
the lenticular and of the covert racism of color blindness [were]
ported into our computational systems."[87] McPherson feels that
although the Unix hackers and the FOSS revolutionaries who
worked in their image were consciously committed to racial
equality, they unwittingly incorporated modes of segregationist
thinking into the operating system that parallel the ones under-
girding racial tension in the United States today, despite American society's having ostensibly entered a postracial age.

McPherson's arguments constitute her subjective philosophical analysis of Unix design rather than any objective evidence of conscious exclusion by designers. In focusing on Unix,
her work also does not provide evidence of how segregationist thinking might later have affected FOSS projects properly
defined, such as GNU or Linux. Programmers and computer
users are unlikely to have chosen not to embrace FOSS because
they believed that Unix or other technologies important to the

FOSS community are subject to a nefarious subconscious disparaging of racial or gender equality. Still, McPherson's interpretation is one of the few that have tried to think in the broadest sense about why the FOSS community remains markedly white and male. She provides an interesting answer. One hopes that future researchers will expand on her intriguing work.

## CONTINUING THE FOSS REVOLUTION

As this chapter has shown, the FOSS community has struggled to define its center over the past decade. Android and Ubuntu are successful when measured in terms of market penetration but have been perceived as failures by hackers who do not see these initiatives as fulfillments of the true promise and potential of the FOSS revolution. Meanwhile, even as FOSS principles have exerted an ever-greater influence on society as a whole, the FOSS community has grappled internally with issues of inclusion and exclusion.

Against this backdrop, it is hard to argue that the FOSS revolution is over. Instead, like the major political revolutions that preceded it, the FOSS revolution has entered a phase in which the meanings and end goals of the FOSS movement are subject to continual debate and reinterpretation within different strands of the FOSS community.

Various FOSS factions have now learned to get along. The self-described pragmatists associated with Linux and open source have ceased to battle the Free Software Foundation's supporters on a continual basis. And FOSS no longer faces existential external threats from companies like Microsoft, which is now running and distributing FOSS software of its own. In these areas, the FOSS revolution has reached a state of equilib-rium. The wild upheavals of previous decades have passed.

Yet that does not mean that all FOSS hackers have been fully satisfied or feel prepared to meet the demands that the future will bring as the software world keeps evolving. Until they do, the FOSS revolution will continue. To declare it over now would be just as shortsighted as Napoleon's attempt to end the French Revolution by dictate two centuries ago.

# Notes

## FOREWORD

1. https://cyber.harvard.edu/property00/respect/fisher.html.
2. https://en.wikipedia.org/wiki/Clarke%27s_three_laws.

## INTRODUCTION

1. Rousseau, *Social Contract*, 165.
2. Baker and Edelstein, *Scripting Revolution*.
3. Russell, *Open Standards*.
4. Cf. "IBM Considers Free Distribution of DB2," *LXer*, Forum, November 21, 2005, http://lxer.com/module/forums/t/20141.
5. These essays first appeared online and were later published in DiBona, Ockman, and Stone, *Open Sources*. They also are available on Raymond's website at http://www.catb.org/esr/writings/homesteading. References below are to the essays on Raymond's site.
6. Williams, *Free as in Freedom*; Torvalds and Diamond, *Just for Fun*.
7. Salus, *Quarter Century of UNIX*. See also Michael S. Mahoney's invaluable interviews with Unix programmers at http://www.princeton.edu/~hos/Mahoney/unixhistory.
8. Bretthauer, "Open Source Software." Bretthauer describes his interest in free software at the time in an email dated March 23, 2016.
9. Ensmenger, "Open Source's Lessons," 104.
10. Ibid., 102.
11. Ibid.

12. Ensmenger, *Computer Boys*.

13. Weber, *Success*, 1.

14. Ibid., 2.

15. Kelty, *Two Bits*.

16. Ibid., 7.

17. Russell, *Open Standards*.

18. Yood, "History of Computing," 88.

19. Chun, *Programmed Visions*; Manovich, *Software Takes Command*; Frabetti, *Software Theory*; Fuller, *Behind the Blip*; Berry, *Philosophy of Software*; von Krogh and von Hippel, "Promise of Research."

20. Manovich, *Software Takes Command*.

21. Chun, *Programmed Visions*, 24.

22. Rich Sands, "Open Source by the Numbers," *SlideShare*, April 6, 2012, http://www.slideshare.net/blackducksoftware/open-source-by-the-numbers.

23. Michele Chubirka, "Open-Source vs. Commercial Software: A False Dilemma," *InformationWeek*, May 14, 2014, http://www.informationweek .com/strategic-cio/it-strategy/open-source-vs-commercial-software-a-false -dilemma/d/d-id/1252665.

24. Torvalds and Diamond, *Just for Fun*; Raymond, "Revenge."

25. Ralph Waldo Emerson, "Concord Hymn," *Poetry Foundation*, accessed April 27, 2016, https://www.poetryfoundation.org/poems-and-poets/poems/ detail/45870; Raymond, "Revenge."

26. *Revolution OS*.

27. DiBona, Ockman, and Stone, *Open Sources*, 60–70.

28. See also Mark Fidelman, "The Fourteen Key Events That Led to a Free and Open Source Software (FOSS) Revolution," *CloudAve*, April 19, 2010, https://www.cloudave.com/503/the-14-key-events-that-led-to-a-free-and -open-source-software-foss-revolution.

29. On the meaning of the "revolutionary script," see Baker and Edelstein, *Scripting Revolution*.

30. Darnton, *Literary Underground*, 21.

## CHAPTER 1: THE PATH TO REVOLUTION

1. Edelstein, *The Terror of Natural Right*, 12.

2. Raymond, "Cathedral."

3. Salus, *Quarter Century of UNIX*, 7.

4. Ibid., 190.

5. Cf. Torvalds and Diamond, *Just for Fun*, 56. On the Spacewar game, which also originated in the 1960s, see Levy, *Hackers*, 57–77.

6. Salus, *Quarter Century of UNIX*, 5.

7. Corbató, Saltzer, and Clingen, "Multics."

8. Ritchie, "Evolution." Ritchie's history of Unix, which he published in an AT&T technical journal in 1984, was adapted from a paper presented at the Language Design and Programming Methodology conference in Sydney, Australia, in September 1979.

9. Ritchie, "Evolution."

10. Salus, *Quarter Century of UNIX*, 9; Ritchie, "Evolution."

11. Ritchie, "Evolution."

12. McIlroy, "Research UNIX Reader."

13. "History of Unix," *Byte*.

14. Salus, *Quarter Century of UNIX*, 70–71.

15. Raymond, "Cathedral."

16. Quoted in Salus, *Quarter Century of UNIX*, 138.

17. Ibid., 142.

18. Ibid., 68–69.

19. "Armando Stretter," *Unix Guru Universe*, http://www.ugu.com/sui/ugu/show?I=info.Armando_Stettner; Eric S. Raymond, "Live Free or Die!," *The Jargon File*, http://catb.org/~esr/jargon/html/L/Live-Free-Or-Die-.html.

20. Raymond, "Brief History"; Salus, *Quarter Century of UNIX*, 190.

21. Quoted in Salus, *Quarter Century of UNIX*, 135.

22. For an overview of the reverse-engineering process, see Ben Everard, "Drive It Yourself: USB Car," *Linux Voice*, March 20, 2015, https://www.linuxvoice.com/drive-it-yourself-usb-car-6.

23. Salus, *Quarter Century of UNIX*, 65.

24. Ibid., 56–57, 65.

25. Ibid., 189–90.

26. Ibid., 190, 222.

27. Levy, *Hackers*, 9–10.

28. Subsequent, revised editions of *Hackers* appeared in 1994 and 2010.

29. Levy, *Hackers*, 8–12.

30. Ibid., 95.

31. Ibid., 10.

32. Ibid., 27.

33. Ibid., 28–34.

34. Regarding Levy's description of Stallman as the "last true hacker," Stallman told me that "There is a big misunderstanding about that phrase. Levy called a certain community of hackers 'the true hackers.' Other communities he called 'the microcomputer hackers' and 'the game hackers.' We who were in the first of these communities never used those terms; we never claimed that only we were truly hackers. When Levy called me 'the last of the true hackers,' he did not mean 'the last real hacker.' Rather, he meant, the last person in the 'true hackers' community who hadn't abandoned it."

35. Raymond, "Brief History."

36. Eric Raymond, "Hacker Ethic," *Jargon File*, http://www.catb.org/jargon/html/H/hacker-ethic.html.

37. Eric Raymond, "Hacker Howto," http://www.catb.org/esr/faqs/hacker-howto.html.

38. "Just Use the GPL," Slashdot, August 24, 2007, https://linux.slashdot.org/story/07/08/24/2017257/foss-license-proliferation-adding-complexity.

39. Himanen, *Hacker Ethic*.

40. Ensmenger, *Computer Boys*.

41. Russell, *Open Standards*.

42. Raymond, "Homesteading."

43. Bezroukov, "Open Source Software."

44. Raymond, "Homesteading."

45. Williams, *Free as in Freedom*, 102.

46. Quoted in Salus, *Quarter Century of UNIX*, 7.

47. DiBona, Ockman, and Stone, *Open Sources*, 32.

48. Quoted in Salus, *Quarter Century of UNIX*, 161.

49. Baker, *Inventing the French Revolution*, 204.

**CHAPTER 2: INVENTING THE FOSS REVOLUTION**

1. Baker, *Inventing the French Revolution*, 203–204.

2. Salus, *Quarter Century of UNIX*, 151–152.

3. Ibid., 222.

4. Ibid.

5. Quoted in ibid., 193.

6.  Ibid.
7.  Quoted in ibid., 232.
8.  "Validating the Source Code."
9.  Gillin, "IBM's Object Code."
10. Bradford, "History of Unix on the PC."
11. Salus, *Quarter Century of UNIX*, 222.
12. Marshall Kirk McKusick, "Twenty Years of Berkeley Unix: From AT&T-Owned to Freely Redistributable," in Dibona, Ockman, and Stone, *Open Sources*, 41. Salus, *Quarter Century of UNIX*, 165, identified November 1988 as the month of NET 1's release. I have followed the indications of McKusick because he was personally involved in BSD development.
13. Salus, *Quarter Century of UNIX*, 223.
14. GNU notes on NET 2, https://ftp.gnu.org/non-gnu/net2-bsd.README.
15. Salus, *Quarter Century of UNIX*, 190, 222–223.
16. Ibid., 165.
17. Ibid., 217–218.
18. *GNU's Bulletin*, June 1988.
19. Ibid., June 1989.
20. Salus, *Quarter Century of UNIX*, 218.
21. Williams, *Free as in Freedom*, 29.
22. Steed, "Freedom's Forgotten Prophet.
23. Williams, *Free as in Freedom*, 33.
24. Ibid., 45–47.
25. Weizenbaum, *Computer Power*, 116.
26. Williams, *Free as in Freedom*, 91–92.
27. Ibid., 4–5.
28. Ibid., 7–9.
29. McHugh, "For the Love of Hacking.".
30. Quoted in Williams, *Free as in Freedom*, 11.
31. "New Unix Implementation."
32. "How to Pronounce GNU," http://www.gnu.org/gnu/pronunciation.en.html.
33. Quoted in Williams, *Free as in Freedom*, 103.
34. "New Unix Implementation."
35. Ibid.
36. Ibid.
37. Ibid.

38. Ibid.

39. Raymond, "Brief History."

40. Email with the author, May 7, 2015.

41. Ibid.

42. Williams, *Free as in Freedom*, 103.

43. Quoted in ibid., 105.

44. Ibid., 104.

45. Ibid., 105.

46. Ibid., 106; United States Census Bureau, "Historical Income Tables—Households," accessed April 25, 2016, http://www.census.gov/hhes/www/income/data/historical/household.

47. Email to the author, May 7, 2015.

48. Ibid.

49. *GNU's Bulletin*, February 1986.

50. Ibid.

51. Ibid.

52. Ibid.

53. Ibid., February 1988.

54. Ibid., June 1987.

55. Ibid., February 1988.

56. Ibid., January 1989.

57. Ibid., June 1995.

58. Ibid., February 1986.

59. Ibid., June 1987.

60. Michael Tiemann, "Future of Cygnus Solutions: An Entrepreneur's Account," in DiBona, Ockman, and Stone, *Open Sources*, 71.

61. *GNU's Bulletin*, June 1991; ibid., January 1992.

62. Ibid., June 1989.

63. Ibid., June 1991; ibid., January 1993.

64. Ibid., January 1995.

65. Ibid., February 1988.

66. Ibid.; ibid., June 1988.

67. Williams, *Free as in Freedom*, 106.

68. Email to author, April 27, 2015.

69. *GNU's Bulletin*, June 1987.

70. Ibid., January 1990.

71. Ibid., June 1991; ibid., June 1995.

72. Hall, *Digitize*, 206.

73. "New Unix Implementation."

74. *GNU's Bulletin*, June 1987.

75. Lemley, *Software and Internet Law*, 34–35.

76. Williams, *Free as in Freedom*, 124.

77. Ibid., 125.

78. Ibid., 128.

79. Ibid., 126.

80. *GNU's Bulletin*, June 1988.

81. Ibid., June 1989.

82. Ibid., January 1989.

83. "Various Licenses and Comments about Them," https://www.gnu.org/licenses/license-list.en.html; Black Duck Software, "Top Twenty Open Source Licenses," https://www.blackducksoftware.com/resources/data/top-20-open-source-licenses.

84. *GNU's Bulletin*, January 1987.

85. Ibid., June 1989; ibid., January 1990; ibid., January 1991.

86. Ibid., June 1988; ibid., January 1995.

87. Ibid., January 1992.

88. Ibid., June 1988.

89. Ibid., January 1990.

90. "New Unix Implementation."

91. *GNU's Bulletin*, February 1986.

92. Ibid., January 1987.

93. Ibid., June 1987.

94. Ibid., June 1988; ibid., January 1989.

95. Ibid., June 1989.

96. Ibid., January 1989.

97. Ibid., June 1991.

98. Richard Stallman, "The GNU Operating System and the Free Software Movement," in DiBona, Ockman, and Stone, *Open Sources*, 65.

99. Torvalds, "The Linux Edge," in DiBona, Ockman, and Stone, *Open Sources*, 103. Torvalds admitted in 1992 that "microkernels are nicer" in theory, but he did not think them practical for real-world applications. "The Tanenbaum-Torvalds Debate," in Dibona, Ockman, and Stone, *Open Sources*, 224.

100. *GNU's Bulletin*, June 1992.

101. Ibid., January 1994.

102. Ibid., June 1992.

103. Ibid., January 1993.

104. Ibid., January 1995.

105. Ibid., January 1996.

106. Ibid., July 1996.

107. "RMS AMA," July 29, 2010, https://www.redditblog.com/2010/07/rms
-ama.html.

108. Greenwood, "Why Is Linux Successful?"

109. *GNU's Bulletin*, March 1998.

110. "Porting the Hurd to Another Microkernel," http://www.gnu.org/software/
hurd/history/port_to_another_microkernel.html.

111. Email to author, May 8, 2015.

112. Ibid.

113. Ibid.

114. *GNU's Bulletin*, June 1995.

115. Reuven M. Lerner, "Stallman Wins $240,000 in MacArthur Award," *The
Tech*, July 18, 1990, http://tech.mit.edu/V110/N30/rms.30n.html.

116. Van Rossum, "Origin of BDFL."

117. Email to author, May 8, 2015.

118. Raymond, "Cathedral."

119. Brooks, *Mythical Man-Month*, 13-14; Frabetti, *Software Theory*, 100.

120. *GNU's Bulletin*, January 1997.

121. Ibid., February 1986.

122. Ibid., January 1992.

123. Ibid., January 1993.

**CHAPTER 3: A KERNEL OF HOPE**

1. Garfinkel, "Is Stallman Stalled?"

2. Quoted in Williams, *Free as in Freedom*, 145.

3. Salus, *Quarter Century of UNIX*, 223.

4. *GNU's Bulletin*, June 1992.

5. McKusick, "Twenty Years," in DiBona, Ockman, and Stone, *Open Sources*,
44.

6. UNIX System Laboratories and Regents of the University of California, "Settlement Agreement," Bell Laboratories, https://www.bell-labs.com/usr/dmr/www/bsdi/USLsettlement.pdf.

7. McKusick, "Twenty Years," in DiBona, Ockman, and Stone, *Open Sources*, 46.

8. Ibid., 46.

9. Torvalds and Diamond, *Just for Fun*, 39.

10. Ibid., 41.

11. Ibid., 51.

12. Ibid., 53.

13. Paju, "National Projects."

14. Young, "Interview."

15. Paju, "National Projects," 84.

16. Torvalds and Diamond, *Just for Fun*, 94.

17. Tanenbaum, *Operating Systems*.

18. Torvalds and Diamond, *Just for Fun*, 51–52.

19. Ibid., 52.

20. Tanenbaum, "Some Notes."

21. Ibid.

22. Torvalds and Diamond, *Just for Fun*, 60–61.

23. Ibid., 62.

24. Ibid.

25. Tanenbaum and Torvalds, et al., "LINUX Is Obsolete," January 31, 1992.

26. Ibid.

27. Tanenbaum and Torvalds, et al., "LINUX Is Obsolete," January 29, 1992.

28. Torvalds and Diamond, *Just for Fun*, 61.

29. Email to author, May 6, 2016.

30. Tanenbaum, "Some Notes."

31. Ibid.

32. Torvalds and Diamond, *Just for Fun*, 58.

33. Torvalds, "LINUX: A Free Unix-386 Kernel."

34. Torvalds and Diamond, *Just for Fun*, 58, 194.

35. Ibid., 73.

36. https://twitter.com/linus__torvalds/status/296333253571387392, January 29, 2013, accessed August 18, 2016.

37. Torvalds, "What Would You Like to See Most in Minix?"

38.  Torvalds and Diamond, *Just for Fun*, 94.

39.  Ibid., 62–63.

40.  Ibid., 77–78.

41.  Ibid., 78.

42.  Torvalds, "LINUX's History."

43.  Torvalds and Diamond, *Just for Fun*, 77–78.

44.  Torvalds, "LINUX's History."

45.  Torvalds and Diamond, *Just for Fun*, 79–80.

46.  Torvalds, "LINUX's History."

47.  Ibid.

48.  Ibid.

49.  Ibid.

50.  Ibid.

51.  Ibid.

52.  Torvalds, "What Would You Like to See Most in Minix?"

53.  Torvalds and Diamond, *Just for Fun*, 81.

54.  Lovell, "And Then Came?"

55.  Tanenbaum, "Some Notes."

56.  *GNU's Bulletin*, March 1998.

57.  Torvalds and Diamond, *Just for Fun*, 87.

58.  Ibid., 84.

59.  Ibid., 87.

60.  Ibid., 90.

61.  Torvalds, "Free Minix-like Kernel Sources."

62.  Quoted in Torvalds, "LINUX's History."

63.  Torvalds, "Free Minix-like Kernel Sources."

64.  Torvalds and Diamond, *Just for Fun*, 90.

65.  Ibid., 91.

66.  Ibid., 93.

67.  Ibid.

68.  Ibid.

69.  Torvalds and Diamond, *Just for Fun*, 94.

70.  Tanenbaum and Torvalds, et al., "LINUX Is Obsolete."

71.  Ibid.

72.  Torvalds and Diamond, *Just for Fun*, 111.

73.  Tanenbaum and Torvalds, et al., "LINUX Is Obsolete."

74.  Torvalds and Diamond, *Just for Fun*, 112.

75. Torvalds, "LINUX: A Free Unix-386 Kernel."

76. Torvalds, "LINUX's History."

77. Torvalds and Diamond, *Just for Fun*, 92.

78. Torvalds, "LINUX's History."

79. Torvalds and Diamond, *Just for Fun*, 117.

80. Ibid., 116–117.

81. Wirzenius, "Linux at Twenty."

82. Young, "Interview with Linus."

83. Torvalds and Diamond, *Just for Fun*, 115.

84. Ibid., 127.

85. Ibid.

86. *GNU's Bulletin*, January 1994.

87. Young, "Interview with Linus."

88. Hillesley, "Asterix, the Gall."

89. Torvalds and Diamond, *Just for Fun*, 133.

90. Marjorie Richardson, "Ownership of Linux Trademark Resolved," *Linux Journal*, November 1, 1997; Torvalds and Diamond, *Just for Fun*, 134.

91. Quoted in Hillesley, "Asterix, the Gall."

92. Ibid.

93. Torvalds, "Notes for Linux Release 0.01."

94. Email to author, May 6, 2016.

95. Ibid.

96. Torvalds and Diamond, *Just for Fun*, 95.

97. Ibid., 95–96; Torvalds, "Release Notes for Linux v0.12."

98. Torvalds, "LINUX's History."

99. Torvalds and Diamond, *Just for Fun*, 96.

100. Ibid., 97.

101. Young, "Interview with Linus."

102. Torvalds and Diamond, *Just for Fun*, 96–97.

103. Quoted in Williams, *Free as in Freedom*, 144.

104. Quoted in ibid., 145.

105. Quoted in ibid.

106. *GNU's Bulletin*, June 1992.

107. Ibid., June 1994.

108. Ibid, June 1994.

109. Ibid., January 1995; Torvalds and Diamond, *Just for Fun*, 132.

110. *GNU's Bulletin*, January 1996.

111. "Brief History of Debian."

112. Quoted in Williams, *Free as in Freedom*, 150.

113. *GNU's Bulletin*, July 1996.

114. Williams, *Free as in Freedom*, 148.

115. Ibid.

116. *GNU's Bulletin*, July 1996.

117. Ibid., July 1997.

118. Ibid., March 1998.

119. Kahney, "Linux's Forgotten Man."

120. Salus, *Quarter Century of UNIX*, 210, 223.

121. Wirzenius, "Linux Anecdotes."

122. Tanenbaum, "Some Notes."

123. McKusick, "Twenty Years," in DiBona, Ockman, and Stone, *Open Sources*, 42.

124. Ibid., 45.

125. Torvalds, "LINUX: A Free Unix-386 Kernel."

126. Quoted in Torvalds, "LINUX's History."

127. McKusick, "Twenty Years," in DiBona, Ockman, and Stone, *Open Sources*, 42.

128. Tanenbaum, "Some Notes."

129. Email to author, May 6, 2016.

130. Ibid.

131. Young, "Interview with Linus."

## CHAPTER 4: THE MODERATE FOSS REVOLUTION

1.  Berlich, "Early History."

2.  Ibid.

3.  Salus, "Daemon."

4.  Slackware 2.1 README, accessed April 26, 2016, https://web.archive.org/web/20111009004917/http://ftp.df.lth.se/pub/slackware/slackware-2.1/README.210.

5.  Hameleers, "History of Slackware."

6.  Salus, "Daemon."

7.  Young, "Giving It Away: How Red Hat Software Stumbled across a New Economic Model and Helped Improve an Industry," in DiBona, Ockman, and Stone, *Open Sources*, 113.

8. "Brief History of Debian."

9. Ibid.

10. Ibid.

11. Perens, "Debian Linux."

12. Ibid.

13. Chris Williams, "Debian Founder Ian Murdock Killed Himself," *The Register*, July 7, 2016, accessed December 20, 2016, www.theregister.co.uk/2016/07/07/ian_murdock_autopsy.

14. Ingo, *Open Life*.

15. Ibid.

16. John Dvorak, "The Lindows Conundrum," *PC Magazine*, October 26, 2001.

17. Bloomberg News, "Technology Briefing. Software: Lindows and Microsoft Settle Suit," *New York Times*, July 20, 2004.

18. Bmasnick, "Review: Mandrake Linux 9.0," *ExtremeTech*, October 21, 2002, accessed July 21, 2015, http://www.extremetech.com/computing/52240-review-mandrake-linux-90.

19. Michael, "Mandrake Appealing to Community, Again," Slashdot, December 20, 2002, accessed April 26, 2016, https://linux.slashdot.org/story/02/12/20/1815214/mandrake-appealing-to-community-again.

20. Julie Bort, "A Linux Company That Spent Seventeen Years Competing with Windows Is Officially Over," *Business Insider*, May 26, 2015, accessed April 26, 2016, http://finance.yahoo.com/news/linux-company-spent-17-years-171430998.html.

21. "How *Not* to Fork Gentoo Linux," *DistroWatch Weekly*, June 30, 2003, accessed April 26, 2016, https://distrowatch.com/weekly.php?issue=20030630.

22. Berlich, "Early History."

23. "How to Build Your Own Linux Distro," http://www.wikihow.com/Build-Your-Own-Linux-Distribution.

24. McKusick, "Twenty Years," in DiBona, Ockman, and Stone, *Open Sources*, 43.

25. Ibid.

26. Ibid.

27. *GNU's Bulletin*, January 1987; Raymond, *Art of Unix*.

28. Robert W. Scheifler, "Debut of X," *Talisman General Information*, June 19, 1984, accessed April 26, 2016, http://www.talisman.org/x-debut.shtml.

29. Jon Brodkin, "Intel Rejection of Ubuntu's Mir Patch Forces Canonical to Go Own Way," *Ars Technica*, September 9, 2013, accessed April 26, 2016, http://arstechnica.com/information-technology/2013/09/intel-rejection-of-ubuntus-mir-patch-forces-canonical-to-go-own-way.

30. "RMS on the GPLing of Qt."

31. Ettrich, "New Project."

32. Richard Stallman, "GNU Operating System and the Free Software Movement," in DiBona, Ockman, and Stone, *Open Sources*, 66–69.

33. Bruce Perens, "The Open Source Definition," in DiBona, Ockman, and Stone, *Open Sources*, 175.

34. Stallman, "The GNU Project."

35. "The GNOME Desktop Project (fwd)," August 28, 1997, accessed April 26, 2016, https://lists.debian.org/debian-user/1997/08/msg02286.html.

36. DiBona, Ockman, and Stone, *Open Sources*, 66–69, 175.

37. "The Q Public License Version 1.0," 1999–2000, accessed April 26, 2016, http://web.archive.org/web/20010911005542/http://doc.trolltech.com/3.0/license.html.

38. Stallman, "Stallman on Qt."

39. Ibid.

40. "RMS on the GPLing of Qt."

41. Matt Heck, "Re: [freeqt] Hello All … ," Harmony mailing list archive, August 4, 1999, accessed April 26, 2016, http://marc.info/?l=kde-freeqt&m=93380255719107&w=2.

42. "GNOME 2.0 Release Notes," accessed April 26, 2016, https://help.gnome.org/misc/release-notes/2.0.

43. These included winGTK (http://wingtk.sourceforge.net) and CyGNOME (http://cygnome.sourceforge.net).

44. Eugenia Loli, "GNOME-Office 1.0 Released; Nautilus Becomes Object-Oriented," *OSnews*, September 15, 2003, accessed April 26, 2016, http://www.osnews.com/comments/4548.

45. "OpenOffice.org History and Milestones 1999–2005," OpenOffice, accessed April 26, 2016, https://www.openoffice.org/about_us/milestones.html; "A Short History of OpenOffice.org," *OooAuthors User Manual*, OpenOffice, accessed April 26, 2016, https://web.archive.org/web/20130918165733/http://wiki.openoffice.org/wiki/Documentation/OOoAuthors_User_Manual/Getting_Started/A_short_history_of_OpenOffice.org; Jack Loftus, "Desktop Apps Ripe Turf for Open Source," *Search Enterprise Linux*, Octo-

ber 4, 2004, accessed April 26, 2016, http://searchenterpriselinux
.techtarget.com/news/1011227/Desktop-apps-ripe-turf-for-open-source.

46. Joe Barr, "Ximian Evolution 1.0 Links Linux to Exchange," *The Register*,
    December 3, 2001, accessed April 26, 2016, http://www.theregister.co.uk/
    2001/12/03/ximian_evolution_1_0_links.

47. "Project: Evolution," accessed April 26, 2016, http://web.archive.org/
    web/20070528161309/http://forge.novell.com/modules/xfcontent/
    downloads.php/evolution/builds/Evolution%202.6%20for%20Mac%20
    OS%20X%20/; Nat Friedman, "Evolution for Windows," Blog, January
    17, 2005, accessed April 26, 2016, http://archive.is/qHUOl.

48. Mitchell Baker, "Thunderbird: Stability and Community Innovation,"
    Blog, July 6, 2012, accessed April 26, 2016, http://blog.lizardwrangler
    .com/2012/07/06/thunderbird-stability-and-community-innovation.

49. "Wine History."

50. Young, "Interview with Linus."

51. "Wine History."

52. Raymond, *Halloween Documents*, "Halloween Document II."

53. "Wine History."

54. DiBona, Ockman, and Stone, *Open Sources*, 15.

55. Tim Smith and François Flückiger, "Licensing the Web," CERN, accessed
    April 26, 2016, http://home.cern/topics/birth-web/licensing-web.

56. *GNU's Bulletin*, June 1991.

57. "Apache Server Frequently Asked Questions," accessed April 26, 2016,
    https://www.bigbiz.com/docs/1.1/1.0/FAQ.html.

58. "About the Apache HTTP Server."

59. "Re: Informing NCSA, Archive of the List," February 28, 1995, accessed
    April 26, 2016, http://mail-archives.apache.org/mod_mbox/httpd-dev/199503
    .mbox/%3C9502281620.AA24455@volterra%3E.

60. Some noncontemporary sources claim that "the name 'Apache' was chosen
    from respect for the Native American Apache Nation" and that the play on
    "a patchy server" was not the name's origin. Cf. http://www.apache.org/
    foundation/how-it-works.html#history. But an email from early in the proj-
    ect's history indicates instead that "If you're wondering about the name, say
    'Apache server' ten time [sic] fast." Robert Thau, "Re: Informing NCSA,
    Archive of the List," February 28, 1995, accessed April 26, 2016,
    http://mail-archives.apache.org/mod_mbox/httpd-dev/199503.mbox/
    %3C9502281620.AA24455@volterra%3E.

61. Robert Thau, "The Political Correctness Question," April 1995, accessed April 26, 2016, http://mail-archives.apache.org/mod_mbox/httpd-dev/199504.mbox/%3C9504230106.AA04957%40volterra%3E.

62. Randy Terbush, "Mission Statement," March 12, 1995, accessed April 26, 2016, http://mail-archives.apache.org/mod_mbox/httpd-dev/199503.mbox/%3C199503121853.MAA07180%40sierra.zyzzyva.com%3E.

63. "About the Apache HTTP Server."

64. "Apache License 1.0," accessed April 27, 2016, https://www.apache.org/licenses/LICENSE-1.0.

65. Rob Hartill, "Apache LICENSE (fwd)," July 5, 1995, accessed April 27, 2016, https://mail-archives.apache.org/mod_mbox/httpd-dev/199507.mbox/%3C199507052206.PAA11787%40taz.hyperreal.com%3E.

66. Brian Tao, "Re: Apache LICENSE (fwd)," July 7, 1995, accessed April 27, 2016, https://mail-archives.apache.org/mod_mbox/httpd-dev/199507.mbox/%3CPine.BSI.3.91.950708001729.14082E-100000%40aries%3E.

67. Randy Terbush, "Re: Apache LICENSE (fwd)," July 6, 1995, accessed April 27, 2016, https://mail-archives.apache.org/mod_mbox/httpd-dev/199507.mbox/raw/%3CPine.BSI.3.91.950706221100.4165r-110000%40taz.hyperreal.com%3E.

68. "Apache License 1.1," accessed April 27, 2016, https://www.apache.org/licenses/LICENSE-1.1.

69. "Various Licenses and Comments about Them," accessed April 27, 2016, https://www.gnu.org/licenses/license-list.html#apache2.

70. "How the ASF Works," accessed April 27, 2016, http://www.apache.org/foundation/how-it-works.html#meritocracy.

71. James Davidson, "NEWS: Jakarta Goes LIVE," October 16, 1999, accessed April 26, 2016, http://mail-archives.apache.org/mod_mbox/jakarta-announcements/199910.mbox/%3C025301bf1824%24dd9e6da0%24a447fea9%40paris%3E.

72. "Report from ApacheCon."

73. "Certificate of Incorporation of the Apache Software Foundation," accessed April 27, 2016, https://web.archive.org/web/20090531160220/http://apache.org/foundation/records/certificate.html; "The Apache Group Incorporates as the Apache Software Foundation," June 30, 1999, accessed April 27, 2016, http://apache.org/foundation/press/pr_1999_06_30.html.

74. "ASF History Goals," accessed April 27, 2016, http://www.apache.org/history/goals.html. This page has not been significantly changed since at

least December 2002, when it was first archived by archive.org. See https://
web.archive.org/web/20021210113807/http://www.apache.org/history/
goals.html.

75. "Projects Directory," accessed April 27, 2016, https://projects.apache.org.

76. "Samba Team Announces Samba 1.9.17," Press release, August 26, 1997, accessed April 27, 2016, https://www.samba.org/samba/history/samba 1.9.17.html.

77. "History of PHP," accessed April 27, 2016, http://php.net/manual/en/ history.php.php.

78. Lauren Orsini, "PHP, Once the Web's Favorite Programming Language, Is on the Wane," *ReadWrite*, August 11, 2014, accessed April 27, 2016, http:// readwrite.com/2014/08/11/why-learn-php.

79. Lutz, *Programming Python*; "perlhist," accessed April 27, 2016, http:// perldoc.perl.org/perlhist.html.

80. Harrison and Feuerstein, *MySQL*, 5–6, 1; Pachev, *Understanding*, 1.

81. Morrison and Snodgrass, "Computer Science."

82. Young, "Giving It Away: How Red Hat Software Stumbled across a New Economic Model and Helped Improve an Industry," in DiBona, Ockman, and Stone, *Open Sources*, 113–114.

83. Young, "Giving It Away," 115.

84. Ibid., 116.

85. Ibid.

86. Om Malik, "Dell Plus Sun Equals VA Research," *Forbes*, May 3, 1999, accessed April 27, 2016, https://web.archive.org/web/20160303180813/ http://www.forbes.com/1999/05/03/feat.html.

87. John Mark Walker, "Ten Years Gone: The VA Linux Systems IPO," *CNET*, December 10, 2009, accessed April 27, 2016, https://www.cnet.com/ news/10-years-gone-the-va-linux-systems-ipo.

88. "SourceForge, Inc. Changes Its Name to Geeknet, Inc.," November 4, 2009, accessed April 27, 2016, https://web.archive.org/web/20130204051709/ http://investors.geek.net/releasedetail.cfm?ReleaseID=521395.

89. Malik, "Dell Plus Sun."

90. Torvalds and Diamond, *Just for Fun*, 157.

91. "Linux: The Era of Open Innovation," IBM, accessed April 27, 2016, http:// www-03.ibm.com/ibm/history/ibm100/us/en/icons/linux; Mike Miller, "Source Release," November 29, 1996, accessed May 31, 2016, https:// marc.info/?l=mysql&m=87602429318864&w=1.

92. Joe Wilcox, "IBM to Spend $1 Billion on Linux in 2001," *CNET*, December 12, 2000, accessed April 27, 2016, http://www.cnet.com/2100-1001 -249750.html.

93. Torvalds and Diamond, *Just for Fun*, 158.

94. IBM announced another billion-dollar investment in Linux in 2013. Steven Vaughan-Nichols, "IBM and Linux: The Next Billion Dollars," *ZDNet*, September 17, 2013, accessed April 27, 2016, http://www.zdnet .com/article/ibm-and-linux-the-next-billion-dollars.

95. *GNU's Bulletin*, June 1988, January 1995.

96. Torvalds and Diamond, *Just for Fun*, 149.

97. Ibid., 150.

98. Ibid., 151.

99. Ibid.

100. Richard Stallman, "Political Notes from 2011: July–October," October 6, 2011, accessed April 27, 2016, https://stallman.org/archives/2011-jul-oct .html#06_October_2011_%28Steve_Jobs%29.

101. Ibid.

102. Ted Samson, "Richard Stallman, Unrepentant: 'Apple Is Your Enemy,'" *InfoWorld*, January 8, 2013, accessed April 27, 2016, http://www.infoworld .com/article/2616420/techology-business/richard-stallman-unrepentant -apple-is-your-enemy.html.

**CHAPTER 5: THE FOSS REVOLUTIONARY WARS**

1. Raymond, "Homesteading."

2. Mizrach, "Is there a Hacker Ethic."

3. "New Unix Implementation."

4. "The GNU Manifesto," accessed April 27, 2016, http://www.gnu.org/gnu/ manifesto.en.html.

5. *GNU's Bulletin*, July 1996.

6. Kahney, "Forgotten Man."

7. Williams, *Free as in Freedom*, 148–149.

8. Rick Moen, "Fear of Forking," November 1999, accessed April 27, 2016, http://linuxmafia.com/faq/Licensing_and_Law/forking.html.

9. Raymond, "Open Source Summit."

10. Williams, *Free as in Freedom*, 162–163.

11. Ibid.
12. Eric Raymond, "Goodbye, 'Free Software'; Hello, 'Open Source,'" http://www.catb.org/~esr/open-source.html; Williams, *Free as in Freedom*, 164.
13. Raymond, "Revenge."
14. Williams, *Free as in Freedom*, 165.
15. Kahney, "Forgotten Man."
16. Williams, *Free as in Freedom*, 166.
17. Ibid., 115.
18. Torvalds and Diamond, *Just for Fun*, 194–195.
19. Raymond, "Homesteading."
20. Williams, *Free as in Freedom*, 117.
21. Kahney, "Forgotten Man."
22. Email to author, April 27, 2015.
23. "*Reader's Digest* European of the Year: Linus," *Linux.com*, January 20, 2001, accessed April 27, 2016, https://www.linux.com/articles/7109.
24. "Netscape Announces Plans to Make Next-Generation Communicator Source Code Available Free on the Net," January 22, 1998, accessed April 27, 2016, http://web.archive.org/web/20021001071727/wp.netscape.com/newsref/pr/newsrelease558.html.
25. Raymond, "Cathedral."
26. Raymond, "Revenge."
27. Jim Hamerly and Tom Paquin, "Freeing the *Source:* The Story of Mozilla," in DiBona, Ockman, and Stone, *Open Sources*, 197.
28. Ibid., 198–200.
29. Ibid., 200–201.
30. Ibid., 198–202.
31. Torvalds and Diamond, *Just for Fun*, 156.
32. Ibid.
33. Raymond, "Revenge."
34. Raymond, "Cathedral," 62.
35. Raymond, "Revenge."
36. "Linux Sucks!! Long Live Windows," February 5, 1999, alt.comp.linux.xxx.
37. Raymond, "Revenge."
38. Justice.gov, accessed April 27, 2016, https://www.justice.gov/atr/cases/f2600/v-a.pdf.
39. Raymond, *Halloween Documents*, "Halloween Document I."
40. Ibid.

41. Ibid., "Halloween Document II."

42. Ibid., "Halloween Document III."

43. "Microsoft Responds to the Open Source Memo Regarding the Open Source Model and Linux," November 5, 1998, accessed April 27, 2016, http://web.archive.org/web/19991013112307/http://microsoft.com/ntserver/nts/news/mwarv/linuxresp.asp.

44. Stallman, "The GNU Operating System," in DiBona, Ockman, and Stone, *Open Sources*, 70.

45. Perens, "Open Source Definition," 186.

46. Raymond, "Response to Nikolai Bezroukov," accessed April 27, 2016, http://www.catb.org/esr/writings/response-to-bezroukov.html; Williams, *Free as in Freedom*, 15.

47. "Speech Transcript—Craig Mundie, The New York University Stern School of Business," Microsoft.com, May 3, 2001, accessed April 27, 2016, http://news.microsoft.com/speeches/speech-transcript-craig-mundie-the-new-york-university-stern-school-of-business.

48. Dave Newbart, "Microsoft CEO Takes Launch Break with the *Sun-Times*," *Chicago Sun-Times*, June 1, 2001, accessed April 26, 2016, https://web.archive.org/web/20011115003306/http://www.suntimes.com/output/tech/cst-fin-micro01.html.

49. Ibid.

50. Raymond, *Halloween Documents*, "Halloween VII."

51. John Markoff, "Judge Says Unix Copyrights Rightfully Belong to Novell," *New York Times*, August 11, 2007, accessed April 27, 2016, http://www.nytimes.com/2007/08/11/technology/11novell.html.

52. Raymond, *Halloween Documents*, "Halloween X."

53. Markoff, "Judge Says."

54. Kenneth Brown, "*Samizdat*: And Other Issues Regarding the 'Source' of Open-Source Code," May 20, 2004, accessed April 27, 2016, http://www.angelfire.com/linux/toussaint/samizdat/samizdat.pdf.

55. Tanenbaum, "Some Notes."

56. Ibid.

57. Robert Lemos, "Linux Makes a Run for Government," *CNET*, August 16, 2002, accessed April 27, 2016, http://www.cnet.com/Linux+makes+a+run+for+government/2100-1001_3-950083.html.

58. "Dennis Ritchie's Interview for *Samizdat*," *Groklaw*, June 1, 2004, accessed April 27, 2016, http://www.groklaw.net/article.php?story =20040601212559558; "Stallman and Salus Also Contradict Ken Brown's Discredited 'Samizdat,'" *Groklaw*, May 29, 2004, accessed April 27, 2016, http://www.groklaw.net/articlebasic.php?story=20040529153027629.

59. "Microsoft Calls AdTI 'Study' an 'Unhelpful Distraction,'" *Groklaw*, June 14, 2004, accessed April 27, 2016, http://www.groklaw.net/articlebasic .php?story=20040614232501302.

## CHAPTER 6: ENDING THE FOSS REVOLUTION?

1. "Linus Torvalds Talks Future of Linux," *APC*, August 22, 2007, accessed April 27, 2016, http://apcmag.com/linus_torvalds_talks_future_of_linux _page_3.htm/; "'Tux' the Aussie Penguin," *Linux Australia*, accessed May 9, 2016, http://web.archive.org/web/20060507115127/http://www.linux.org .au/linux/tux.

2. Dave Newbart, "Microsoft CEO Takes Launch Break with the *Sun-Times*," *Chicago Sun-Times*, June 1, 2001, accessed April 26, 2016, https://web .archive.org/web/20011115003306/http://www.suntimes.com/output/tech/ cst-fin-micro01.html.

3. "Who Has Faster Pipes? Linux, Win2000, WinXP Compared," Slashdot, October 3, 2001, accessed April 27, 2016, https://developers.slashdot .org/story/01/10/03/176257/who-has-faster-pipes-linux-win2000-winxp -compared.

4. Weber, *Success*.

5. Black Duck and North Bridge, "The Ninth Annual Future of Open Source Survey," 2015, accessed April 27, 2016, https://web.archive.org/web/201 50817010126/https://www.blackducksoftware.com/future-of-open-source.

6. "Dell to Use Ubuntu on Linux PCs," *BBC News*, May 1, 2007, accessed April 27, 2016, http://news.bbc.co.uk/2/hi/business/6610901.stm.

7. "Announcing the Chromium OS Open Source Project," Google Chrome Blog, November 19, 2009, accessed April 27, 2016, https://chrome .blogspot.com/2009/11/announcing-chromium-os-open-source.html.

8. Neil McAllister, "Open Source? HP Enterprise Will Be All-in, Post Split, Says CTO," *The Register*, June 4, 2015, accessed April 26, 2017, http:// www.theregister.co.uk/2015/06/04/hp_enterprise_loves_open_source.

9.    Humayun Shahid, "The Odd Couple: Microsoft and Ubuntu Work It Out in the Cloud," June 14, 2012, accessed April 27, 2016, http://cloudtweaks .com/2012/06/the-odd-couple-microsoft-and-ubuntu-work-it-out-in-the -cloud; "IoT World: Canonical and Industry Leaders Drive IoT Commercialization," Ubuntu, May 11, 2015, https://insights.ubuntu.com/2015/ 05/11/iot-world-canonical-industry-leaders-drive-iot-commercialization/; Mark Shuttleworth, "Comment 1834 for Bug 1," May 30, 2013, accessed April 27, 2016, https://bugs.launchpad.net/ubuntu/+bug/1/comments/1834.

10.   Neil McAllister, "Redmond Top Man Satya Nadella: 'Microsoft LOVES Linux,'" *The Register*, October 20, 2014, accessed April 27, 2016, http:// www.theregister.co.uk/2014/10/20/microsoft_cloud_event.

11.   Cade Metz, "Microsoft Built Its Own Linux Because Everyone Else Did," *Wired*, September 29, 2015, https://www.wired.com/2015/09/microsoft -built-linux-everyone-else.

12.   Steven Vaughan-Nichols, "Microsoft's Love Affair with Linux Deepens," *ZDNet*, September 21, 2015, Comment, accessed April 27, 2016, http:// www.zdnet.com/article/microsoft-the-linux-company.

13.   Roy Schestowitz, "The 'Microsoft Loves Linux' Baloney Is Still Being Floated in the Media," *Techrights*, October 1, 2015, accessed April 27, 2016, http://techrights.org/2015/10/01/microsoft-loves-linux-brainwash.

14.   Steven Vaughan-Nichols, "Ubuntu (Not Linux) on Windows: How It Works," *ZDNet*, March 30, 2016, accessed April 27, 2016, http://www .zdnet.com/article/ubuntu-not-linux-on-windows-how-it-works.

15.   Hall, *Digitize*, 120.

16.   Steve Kovach, "How Android Grew to Be More Popular Than the iPhone," *Business Insider*, August 13, 2013, accessed April 27, 2016, http://www .businessinsider.com/history-of-android-2013-8.

17.   Ben Elgin, "Google Buys Android for Its Mobile Arsenal," *Bloomberg Businessweek*, August 17, 2005, accessed April 27, 2016, https://web.archive.org/ web/20060203184218/http://www.businessweek.com/technology/content/ aug2005/tc20050817_0949_tc024.htm.

18.   "Industry Leaders Announce Open Platform for Mobile Devices," *Open Handset Alliance*, November 5, 2007, accessed April 27, 2016, http://www .openhandsetalliance.com/press_110507.html.

19.   Kovach, "How Android Grew."

20.   Arjun Kharpal, "Google Android Hits Market Share Record with Nearly 9 in Every 10 Smartphones Using It," *CNBC*, November 3, 2016, accessed

December 21, 2016, http://www.cnbc.com/2016/11/03/google-android
-hits-market-share-record-with-nearly-9-in-every-10-smartphones-using-it
.html.

21. "January 2015 Web Server Survey," *Netcraft*, January 15, 2015, accessed
    April 27, 2016, https://news.netcraft.com/archives/2015/01/15/january
    -2015-web-server-survey.html.

22. Katherine Noyes, "If Desktop Linux Is Dead, Someone Had Better Tell All
    Those Users," *PCWorld*, March 26, 2012, accessed April 27, 2016, www
    .pcworld.com/article/252552/if_desktop_linux_is_dead_someone_had
    _better_tell_all_those_users.html.

23. "An Introduction to Android," *SlideShare*, February 19, 2009, accessed
    April 27, 2016, http://www.slideshare.net/natdefreitas/an-introduction-to
    -android.

24. Mathieu Devos, "Bionic vs. glibc Report," Master's thesis, Massachusetts
    Institute of Technology, 2014, http://irati.eu/wp-content/uploads/2012/07/
    bionic_report.pdf.

25. "Industry Leaders Announce Open Platform for Mobile Devices."

26. Ryan Paul, "Why Google Chose the Apache Software License over GPLv2
    for Android," *Ars Technica*, November 6, 2007, accessed April 27, 2016,
    http://arstechnica.com/uncategorized/2007/11/why-google-chose-the-apache
    -software-license-over-gplv2.

27. Ibid.

28. Superglaze, "Android's 'Non-Fragmentation Agreement,'" Slashdot,
    November 13, 2007, accessed April 27, 2016, https://slashdot.org/story/
    07/11/13/1348233/androids-non-fragmentation-agreement.

29. Ibid.

30. Ibid.

31. Ibid.

32. Gavin Clarke, "Stallman: Android Evil, Apple and Microsoft Worse," *The
    Register*, September 20, 2011, accessed April 27, 2016, http://www
    .theregister.co.uk/2011/09/20/stallman_on_android.

33. "Staging: Android: Delete Android Drivers," accessed April 27, 2016, http://
    git.kernel.org/cgit/linux/kernel/git/torvalds/linux.git/commit/?id=b0a0ccfad
    85b3657fe999805df65f5cfe634ab8a.

34. Steven Vaughan-Nichols, "Linus Torvalds on Android, the Linux Fork,"
    *ZDNet*, August 18, 2011, accessed April 27, 2016, http://www.zdnet.com/
    article/linus-torvalds-on-android-the-linux-fork.

35. Ibid.

36. Steven Vaughan-Nichols, "Android Linux FUD Debunked," *ZDNet*, March 22, 2011, accessed April 27, 2016, http://www.zdnet.com/article/android-linux-fud-debunked.

37. Mark Shuttleworth, "Funding Free Software Projects," November 21, 2003, accessed April 27, 2016, http://www.markshuttleworth.com/archives/date/2003/11.

38. "About Kubuntu," Ubuntu, accessed April 27, 2016, https://help.ubuntu.com/kubuntu/desktopguide/C/about-kubuntu.html.

39. Mark Shuttleworth in discussion with the author, June 1, 2015.

40. "What Is Ubuntu?," *Ubuntu Installation Guide*, Ubuntu, accessed April 27, 2016, http://old-releases.ubuntu.com/ubuntu/dists/warty/main/installer-i386/current/doc/manual/en.

41. Mark Shuttleworth in discussion with the author, June 1, 2015.

42. Ibid.

43. Steven Vaughan-Nichols, "Ubuntu Moves Some Linux Development Inside," *ZDNet*, October 19, 2012, accessed April 27, 2016, http://www.zdnet.com/article/ubuntu-moves-some-linux-development-inside.

44. "About," *Ubuntu Insights*, Ubuntu, accessed July 14, 2015, https://insights.ubuntu.com/about.

45. Ashlee Vance, "A Software Populist Who Doesn't Do Windows," *New York Times*, January 10, 2009, accessed April 27, 2016, http://www.nytimes.com/2009/01/11/business/11ubuntu.html; Ryan Paul, "French Police: We Saved Millions of Euros by Adopting Ubuntu," *Ars Technica*, March 11, 2009, accessed April 27, 2016, http://arstechnica.com/information-technology/2009/03/french-police-saves-millions-of-euros-by-adopting-ubuntu.

46. "EC2 Statistics," *The Cloud Market*, accessed April 27, 2016, http://thecloudmarket.com/stats; "Usage Statistics and Market Share of Linux for Websites," *W3Techs*, accessed July 14, 2015, https://w3techs.com/technologies/details/os-linux/all/all.

47. Benjamin Mako Hill, "Announcing Launch of ($10m) Ubuntu Foundation," Ubuntu, July 8, 2005, accessed April 27, 2016, https://lists.ubuntu.com/archives/ubuntu-announce/2005-July/000025.html.

48. Gerry Carr, "ShipIt Comes to an End," Canonical, blog, April 5, 2011, accessed April 27, 2016, http://blog.canonical.com/2011/04/05/shipit-comes-to-an-end.

49. Mark Shuttleworth in discussion with the author, June 1, 2015.

50. Ibid.

51. Ibid.

52. Climenole, "Shuttleworth Answers Ubuntu Linux's Critics," Slashdot, September 14, 2010, accessed April 27, 2016, https://linux.slashdot.org/story/10/09/14/219252/shuttleworth-answers-ubuntu-linuxs-critics.

53. AlexGr, "Is Ubuntu Selling Out or Growing Up?," Slashdot, April 30, 2008, accessed April 27, 2016, https://linux.slashdot.org/story/08/04/30/199204/is-ubuntu-selling-out-or-growing-up.

54. Merrill, "Interview with Linus Torvalds."

55. Richard Stallman, "Ubuntu Spyware: What to Do?," GNU Operating System, accessed April 27, 2016, http://www.gnu.org/philosophy/ubuntu-spyware.en.html; Jon Brodkin, "Richard Stallman Calls Ubuntu 'Spyware' Because It Tracks Searches," *Ars Technica*, December 7, 2012, accessed April 27, 2016, http://arstechnica.com/information-technology/2012/12/richard-stallman-calls-ubuntu-spyware-because-it-tracks-searches/; Timothy, "Ubuntu 14.10 Released with Ambitious Name, But Small Changes," Slashdot, October 23, 2014, accessed April 27, 2016, https://linux.slashdot.org/story/14/10/23/1946243/ubuntu-1410-released-with-ambitious-name-but-small-changes.

56. AlexGr, "Is Ubuntu Selling Out or Growing Up?"

57. Robbie Williamson, "Ubuntu Software Store: What It Does, and How You Can Help," Ubuntu, September 1, 2009, accessed April 27, 2016, https://lists.ubuntu.com/archives/ubuntu-devel/2009-September/028901.html.

58. Matthew East, "Ubuntu Software Store: What It Does, and How You Can Help," Ubuntu, August 27, 2009, accessed April 27, 2016, https://lists.ubuntu.com/archives/ubuntu-devel/2009-August/028814.html.

59. Richard Collins, "Ubuntu's Path to Convergence," Ubuntu, October 20, 2015, accessed April 27, 2016, https://insights.ubuntu.com/2015/10/20/ubuntus-path-to-convergence.

60. "Rackspace Open Sources Cloud Platform," Rackspace, press release, July 19, 2010, accessed April 27, 2016, http://ir.rackspace.com/phoenix.zhtml?c=221673&p=irol-newsArticle&ID=1448761.

61. Angela Bartels, "First OpenStack Release Now Available," Rackspace, blog, October 21, 2010, accessed April 27, 2016, http://blog.rackspace.com/first-openstack-release-now-available.

62. "Openstack Community Contribution in Newton Release," Stackalytics, accessed April 7, 2016, http://stackalytics.com.

63. Cliff Saran, "Is OpenStack Ready for Mass Adoption?," *Computer Weekly*, October 27, 2015, accessed April 27, 2016, http://www.computerweekly .com/news/4500256197/Is-OpenStack-ready-for-mass-adoption; Steven Vaughan-Nichols, "OpenStack Isn't Just Ready for Enterprise Adoptions, It's Already There," *ZDNet*, May 21, 2015, accessed April 27, 2016, www .zdnet.com/article/openstack-isnt-just-ready-for-enterprise-adoption-its -already-there.

64. Bobbie Johnson, "Cloud Computing Is a Trap, Warns GNU Founder Richard Stallman," *The Guardian*, September 29, 2008, accessed April 27, 2016, https://www.theguardian.com/technology/2008/sep/29/cloud.computing .richard.stallman.

65. Richard Stallman, "Who Does That Server Really Serve?," GNU Operating System, accessed April 27, 2016, http://www.gnu.org/philosophy/who-does -that-server-really-serve.en.html.

66. Ibid.

67. "Why the Affero GPL?," GNU Operating System, accessed April 27, 2016, http://www.gnu.org/licenses/why-affero-gpl.html.

68. See the projects Dronecode and Automotive Grade Linux.

69. Bruce Schneir, "The Internet of Things Is Wildly Insecure—and Often Unpatchable," *Wired*, January 6, 2014, accessed April 27, 2016, https:// www.wired.com/2014/01/theres-no-good-way-to-patch-the-internet-of -things-and-thats-a-huge-problem.

70. For example, the b43 FOSS project created its own firmware for use with wireless card components manufactured by Broadcom. See Franceso Gringoli, "Open Source Firmware for Broadcom Wireless Adapters," LWN, blog, January 9, 2009, accessed April 28, 2016, http://lwn.net/Articles/ 314313. On legal issues and reverse engineering, see "Coders' Rights Project Reverse Engineering FAQ," Electronic Frontier Foundation, accessed April 28, 2016, https://www.eff.org/issues/coders/reverse-engineering-faq.

71. Lessig, *Free Culture*, 25.

72. Ibid., xv.

73. Hal Plotkin, "All Hail Creative Commons," *SFGate*, February 11, 2002, accessed April 28, 2016, http://www.sfgate.com/news/article/All-Hail-Creative -Commons-Stanford-professor-2874018.php; "History," Creative Commons, accessed April 28, 2016, https://creativecommons.org/about/history.

74. Richard Stallman, "The Free Universal Encyclopedia and Learning Resource," GNU Operating System, December 18, 2000, accessed April 28, 2016, https://www.gnu.org/encyclopedia/anencyc.txt.

75. Rosenzweig, "Can History Be Open Source?" See also Yochai Benkler, "The New Open-Source Economics," TEDGlobal 2005, transcript, July 2005, accessed April 28, 2016, http://www.ted.com/talks/yochai_benkler_on_the_new_open_source_economics.

76. Ibid.; Benkler, *Wealth of Networks*.

77. Levy, *Hackers*, 31.

78. "Employed Persons by Detailed Occupation," U.S. Bureau of Labor Statistics, accessed April 28, 2016, http://www.bls.gov/cps/cpsaat11.pdf. See also Mahoney, "Boys' Toys and Women's Work."

79. Demby, "Why Isn't Open Source a Gateway for Coders of Color?

80. Loconet, "Open Source Geeks Considered Modern Heroes," Slashdot, November 30, 2004, accessed April 28, 2016, https://developers.slashdot.org/story/04/11/30/176204/open-source-geeks-considered-modern-heroes.

81. Kathryn Zickuhr, "Who's Not Online and Why," Pew Research Center, September 25, 2013, accessed April 28, 2016, http://www.pewinternet.org/2013/09/25/whos-not-online-and-why.

82. Eric Raymond, "War Games II," blog, April 1, 1992, accessed April 28, 2016, http://www.catb.org/esr/writings/wargames.txt.

83. Sam Machkevoch, "Linus Torvalds on Why He Isn't Nice: 'I Don't Care about You,'" *Ars Technica*, January 15, 2015, accessed April 28, 2016, http://arstechnica.com/business/2015/01/linus-torvalds-on-why-he-isnt-nice-i-dont-care-about-you.

84. Raymond, "Cathedral."

85. Weber, *Success*; Kelty, *Two Bits*.

86. "2015 Linux Jobs Report: Linux Professionals in High Demand," Linux Foundation, press release, March 4, 2015, accessed April 28, 2016, https://www.linuxfoundation.org/news-media/announcements/2015/03/2015-linux-jobs-report-linux-professionals-high-demand; Don Willmott, "High Demand Pushes Linux Salaries Higher," *Dice*, February 28, 2013, accessed April 28, 2016, http://insights.dice.com/2013/02/28/demand-for-linux-talent.

87. McPherson, "Why Are the Digital Humanities So White?"

# Glossary

**assembly language**  A type of programming language that gives programmers extensive control over how a computer performs tasks. Because assembly language is specific to a particular type of computer processor, assembly language programs are generally not portable.

**Bell Laboratories**  An AT&T research facility in Murray Hill, New Jersey, where Ken Thompson and Dennis Ritchie developed the first version of Unix in 1969.

**Berkeley Software Distribution (BSD)**  A Unix-like operating system that was developed at the University of California at Berkeley from 1977 until 1995. Begun as an extension of Unix that shared Unix code, it grew into a standalone operating system that was virtually free of Unix code with the 1991 release of Net/2 BSD.

**BSD**  See Berkeley Software Distribution.

**C**  A programming language that was developed by Dennis Ritchie beginning in 1972 to help write Unix.

**closed source software**  Software whose source code is not publicly available.

**cloud computing**  A computing paradigm in which some or all of the data or computing resources of one device are shared over the network with other devices.

**compiler**  A program that translates source code into binary software, or machine code. The GNU Compiler Collection (GCC) compilers are popular among FOSS programmers.

**cross-platform software**  Software that is designed to run on multiple types of operating systems or hardware platforms.

**firmware**   Software that is embedded permanently as part of a hardware device, making it difficult to modify.

**free software**   In the context of the free and open source software movement, free software is defined by two characteristics: first, its source code is publicly available; second, the source code of any derivative works is also publicly available. Free software is often, but not necessarily, free of cost.

**freeware**   Software that can be legally obtained and used free of charge, but whose source code is not publicly available.

**GNU**   1. The project that Richard Stallman announced in 1983 and launched in 1984 for building a Unix-like operating system, including a kernel and user space applications, from original code. 2. The suite of programs that the GNU project produced.

**GNU/Linux**   A generic term for operating systems that use the Linux kernel with programs written by the GNU project. Other programs are included in such systems.

**kernel**   The part of an operating system that performs the basic core functions that are necessary for software to communicate with hardware and for programs to interact with one another.

**library**   A software package that is designed to be shared by multiple programs on the same computer.

**Linux**   An operating system kernel developed under the leadership of Linus Torvalds starting in 1991. Linux provides the kernel for GNU/Linux operating systems.

**live CD**   A compact disk (CD) that can boot a computer and run a standalone system without installing any data to a hard disk. Live CDs (and their more modern cousins, live USBs) allow users to test GNU/Linux distributions easily.

**machine code**   Software that a compiler has prepared for execution by a computer. Machine code can run immediately on the type of system it was compiled for, but because it does not contain source code, it is difficult to study or modify.

**microkernel**   A type of kernel architecture in which the kernel's operations are divided into separate programs to increase modularity. It is the opposite of monolithic kernel design.

**monolithic kernel**  A type of kernel architecture in which all of the basic system functions run as a single kernel program. It is the opposite of microkernel design. Linux is a monolithic kernel.

**Multics**  An operating system collaboratively developed beginning in 1964 by the Massachusetts Institute of Technology, Bell Laboratories, and General Electric. Some of its design principles influenced Unix.

**open source software**  Software whose source code is publicly available. Although this term is sometimes used interchangeably with "free software," some programmers prefer "open source software" because they believe "free software" wrongly implies that software should not cost money. Open source software is also different from free software because some programs that are open source software, but not free software, can be modified without making the source code of the derivative works publicly available.

**patch**  Software that modifies source code in order to update or change a program.

**porting**  The act of adapting software that was written for one type of computer or operating system so that it can run on a different platform.

**POSIX**  A series of standards that define how the components of a Unix-like operating should behave. POSIX compliance ensures compatibility between different types of Unix-like platforms.

**protocol**  A set of regulations that define how software programs should interact with one another or exchange data.

**shell**  The interface between a computer and its user. It may be text-based or graphical. The Bourne Again Shell (Bash) is a popular text-based shell for Unix-like systems.

**source code**  The instructions that a computer uses to perform an action. Source code can be easily read and modified by programmers. A compiler must usually translate source code into a binary before it can be executed.

**Space Travel**  A game that helped inspire Ken Thompson to write the first version of Unix in 1969.

**Unix**  An operating system that was developed beginning in 1969. The first version was written in assembly language at Bell Laboratories, but programmers at many sites around the world contributed code to later versions. It was commercialized by AT&T in 1983. Although the word *Unix* is sometimes used to refer to other operating systems that are designed in the same way as Unix, the latter systems are more properly called Unix-like.

**Unix-like systems**  Operating systems that are designed to function like Unix but do not use Unix code. BSD and GNU/Linux systems are examples of Unix-like operating systems.

**user space**  The part of an operating system in which ordinary programs and applications run. User space is distinct from kernel space, which is controlled by the kernel. Also known as Userland.

**virtual server**  A server that runs as a virtual machine, often as part of a cloud rather than directly on a physical computer.

# Bibliography

## ELECTRONIC ARCHIVES

alt.comp.linux.xxx Usenet Archive. https://archive.org/download/usenet-alt/alt
.comp.linux.xxx.mbox.zip.

Apache Mailing List Archives. http://mail-archives.apache.org/mod_mbox.

comp.os.linux Usenet Archives. https://groups.google.com/forum/#!forum/comp
.os.linux.

comp.os.minix Usenet Archives. https://groups.google.com/forum/#!forum/comp
.os.minix.

*GNU's Bulletin* Online Archive. http://www.gnu.org/bulletins/bulletins.en.html.

MySQL Mailing List Archives. https://marc.info/?l=mysql.

Ubuntu Mailing List Archives. https://lists.ubuntu.com.


## PUBLISHED SOURCES

"About the Apache HTTP Server Project." Apache. Accessed April 26, 2016.
http://httpd.apache.org/ABOUT_APACHE.html.

Baker, Keith. *Inventing the French Revolution: Essays on French Political Culture in
the Eighteenth Century.* Cambridge: Cambridge University Press, 1990.

Baker, Keith, and Dan Edelstein, eds. *Scripting Revolution: A Historical Approach
to the Comparative Study of Revolutions.* Stanford: Stanford University Press, 2015.

Benkler, Yochai. *The Wealth of Networks*. New Haven: Yale University Press, 2006.

Berlich, Ruediger. "The Early History of Linux, Part 2: Re:distribution." *LinuxUser*. April 2001.

Berry, David. *The Philosophy of Software: Code and Mediation in the Digital Age*. New York: Palgrave Macmillan, 2011.

Bezroukov, Nikolai. "Open Source Software Development as a Special Type of Academic Research." *First Monday* (October 1999): 4.

Bradford, Bill. "The History of Unix on the PC: Exploring Lesser-Known Variants." Search Data Center. Accessed June 16, 2016, http://searchdatacenter .techtarget.com/tip/The-history-of-Unix-on-the-PC-Exploring-lesser-known -variants.

Bretthauer, David. "Open Source Software: A History." *UConn Libraries Published Works*. Paper 7. http://digitalcommons.uconn.edu/libr_pubs/7.

"A Brief History of Debian." Debian. Accessed April 26, 2016. https://www .debian.org/doc/manuals/project-history/ch-leaders.en.html.

Brooks, Frederick P. *The Mythical Man-Month and Other Essays on Software Engineering*. Chapel Hill: University of North Carolina at Chapel Hill, 1974.

Campbell-Kelly, Martin. *From Airline Reservations to Sonic the Hedgehog: A History of the Software Industry*. Cambridge, MA: MIT Press, 2003.

Chun, Wendy Hui Kyong. *Programmed Visions: Software and Memory*. Cambridge, MA: MIT Press, 2011.

Corbató, F. J., J. H. Saltzer, and C. T. Clingen. "Multics: The First Seven Years." Paper presented at AFIPS '72: The 1972 Spring Joint Computer Conference. http://www.multicians.org/f7y.html.

Darnton, Robert. *The Literary Underground of the Old Regime*. Cambridge, MA: Harvard University Press, 1982.

Demby, Gene. "Why Isn't Open Source a Gateway for coders of Color?," NPR, December 5, 2013. Accessed April 28, 2016. http://www.npr.org/sections/ codeswitch/2013/12/05/248791579/why-isnt-open-source-a-gateway-for-coders -of-color.

DiBona, Chris, Sam Ockman, and Mark Stone, eds. *Open Sources: Voices from the Open Source Revolution*. Sebastopol, CA: O'Reilly, 1999.

Edelstein, Dan. *The Terror of Natural Right: Republicanism, the Cult of Nature, and the French Revolution*. Chicago: University of Chicago Press, 2009.

Ensmenger, Nathan. *The Computer Boys Take Over: Computers, Programmers and the Politics of Technical Expertise*. Cambridge, MA: MIT Press, 2010.

Ensmenger, Nathan. "Open Source's Lessons for Historians." *IEEE Annals of the History of Computing* 26 (2004): 102–104.

Ettrich, Matthias. "New Project: Kool Desktop Environment. Programmers Wanted!" October 14, 1996. Accessed April 26, 2016. https://groups.google.com/forum/#!msg/de.comp.os.linux.misc/SDbiV3Iat_s/zv_D_2ctS8sJ.

Frabetti, Federica. *Software Theory: A Cultural History*. London: Rowman and Little, 2015.

Fuller, Matthew. *Behind the Blip: Essays on the Culture of Software*. New York: Autonomedia, 2003.

Garfinkel, Simson. "Is Stallman Stalled?" *Wired*, January 1, 1993.

Gillin, Paul. "IBM's Object Code Policy Still Irking Users." *Computerworld* 18 (20) (May 14, 1984).

*GNU's Bulletin*. Various issues. Accessed April 25, 2016. https://www.gnu.org/bulletins/bulletins.html.

Greenwood, Liam. "Why Is Linux Successful? An Opinion." *The Free BSD Diary*, March 20, 1999. Accessed January 20, 2017. http:://www.freebsddiary.org/linux.php.

Hall, Gary. *Digitize This Book! The Politics of New Media, or Why We Need Open Access Now*. Minneapolis: University of Minnesota Press, 2008.

Hameleers, Eric. "A History of Slackware Development." October 2009. Accessed April 26, 2016. http://www.slackware.com/~alien/tdose2009/t-dose-slackware.pdf.

Harrison, Guy, and Steven Feuerstein. *MySQL Stored Procedure Programming*. Sebastopol, CA: O'Reilly, 1996.

Hillesley, Richard. "Asterix, the Gall: The Strange History of Linux and Trademarks." Accessed April 26, 2016. http://tuxdeluxe.org/node/107.

Himanen, Pekka. *The Hacker Ethic: A Radical Approach to the Philosophy of Business*. New York: Random House, 2002.

"The History of Unix." *Byte: The Small Systems Journal* (August 1983): 188.

Ingo, Henrik. *Open Life: The Philosophy of Open Source*. Lulu.com, 2006.

Kahney, Leander. "Linux's Forgotten Man." *Wired*, March 5, 1999.

Kelty, Christopher. *Two Bits: The Cultural Significance of Free Software*. Durham, NC: Duke University Press, 2008.

Lemley, Mark. *Software and Internet Law*. Gaithersburg, MD: Aspen Law and Business, 2000.

Lessig, Lawrence. *Free Culture: How Media Uses Technology and the Law to Lock Down Culture and Control Creativity*. New York: Penguin, 2004.

Levy, Steven. *Hackers: Heroes of the Computer Revolution*. Garden City, NY: Anchor Press, 1984.

Lovell, Anthony. "And Then Came?" Google Groups. Accessed April 26, 2016. https://groups.google.com/forum/#!topic/comp.os.linux/ykCjbjI_Efs.

Lutz, Mark. *Programming Python*. Sebastopol, CA: O'Reilly, 1996.

Lynn, John. *Bayonets of the Republic: Motivation and Tactics in the Army of Revolutionary France, 1791–94*. Urbana: University of Chicago Press, 1984.

Mahoney, Michael S. "Boys' Toys and Women's Work: Feminism Engages Software." In *Feminism in Twentieth-Century Science, Technology and Medicine*, ed. Angela N. H. Creager, Elizabeth Lunbeck, and Londa Schiebinger, 169–185. Chicago: University of Chicago Press, 2001.

Manovich, Lev. *Software Takes Command: Extending the Language of New Media*. London: Bloomsbury, 2013.

McHugh, Josh. "For the Love of Hacking." Forbes, August 10, 1998.

McIlroy, Douglas M. "A Research UNIX Reader: Annotated Excerpts from the Programmer's Manual, 1971–1986." http://www.cs.dartmouth.edu/~doug/reader.pdf.

McPherson, Tara. "Why Are the Digital Humanities So White? Or Thinking the Histories of Race and Computation." In *Debates in the Digital Humanities*, ed. Matthew K. Gold, 139–160. Minneapolis: University of Minnesota Press, 2012.

Merrill, Scott. "An Interview with Linus Torvalds." TechCrunch, April 19, 2012. Accessed April 27, 2016. https://techcrunch.com/2012/04/19/an-interview-with-millenium-technology-prize-finalist-linus-torvalds.

Mizrach, Steven. "Is There a Hacker Ethic for 90s Hackers?" Accessed April 27, 2016. http://www2.fiu.edu/~mizrachs/hackethic.html.

Moody, Glyn. *Rebel Code: Linux and the Open Source Revolution*. London: Penguin, 2002.

Morrison, Clayton T., and Richard T. Snodgrass. "Computer Science Can Use More Science." *Communications of the ACM* 54 (2011): 36–38.

"New Unix Implementation." GNU Operating System. Accessed April 25, 2016. https://www.gnu.org/gnu/initial-announcement.html.

Pachev, Alexander. *Understanding MySQL Internals*. Sebastopol, CA: O'Reilly, 1997.

Paju, Petri. "National Projects and International Users: Finland and Early European Computerization." *Annals of the History of Computing, IEEE* 30 (2008): 77–91.

Perens, Bruce. "Debian Linux Distribution Release 1.1 Now Available." Debian. June 17, 1996. Accessed April 26, 2016. https://lists.debian.org/debian-announce/1996/msg00021.html.

Priestley, Mark. *A Science of Operations: Machines, Logic, and the Invention of Programming*. New York: Springer, 2010.

Raymond, Eric. *The Art of Unix Usability*. April 18, 2004. Accessed April 26, 2016. http://www.catb.org/esr/writings/taouu/html.

Raymond, Eric Steven. "A Brief History of Hackerdom." 1998–2000. Accessed April 25, 2016. http://www.catb.org/esr/writings/homesteading/hacker-history.

Raymond, Eric Steven. "The Cathedral and the Bazaar." Accessed April 25, 2016. http://www.catb.org/esr/writings/homesteading/cathedral-bazaar.

Raymond, Eric. *The Halloween Documents*. Accessed April 26, 2016. http://www.catb.org/esr/halloween/index.html.

Raymond, Eric Steven. "Homesteading the Noosphere." Accessed April 25, 2016. http://www.catb.org/esr/writings/homesteading/homesteading.

Raymond, Eric. "Open Source Summit." *Linux Journal*, June 1, 1998. Accessed April 27, 2016. http://www.linuxjournal.com/article/2918.

Raymond, Eric Steven. "The Revenge of the Hackers." Accessed April 25, 2016. http://www.catb.org/esr/writings/homesteading/hacker-revenge.

"Report from ApacheCon '98." ApacheWeek, October 16, 1998. Accessed April 27, 2016. http://www.apacheweek.com/features/apachecon98.

*Revolution OS.* Directed by J.T.S. Moore. Wonderview Productions, 2001.

Ritchie, Dennis. "The Evolution of the Unix Time-sharing System." *AT&T Bell Laboratories Technical Journal* 63 (October 1984): 1577–1593.

"RMS on the GPLing of Qt and More." Slashdot. September 5, 2000. Accessed April 26, 2016. https://slashdot.org/story/00/09/05/1326250/rms-on-the-gpling -of-qt-and-more.

Rosenzweig, Roy. "Can History Be Open Source? *Wikipedia* and the Future of the Past." *Journal of American History* 98 (June 2006): 117–146.

Rousseau, Jean-Jacques. *The Social Contract and Discourses.* Trans. G. D. H. Cole. New York: Dutton, 1973.

Russell, Andrew. *Open Standards and the Digital Age: History, Ideology, and Networks.* New York: Cambridge University Press, 2014.

Salus, Peter. *A Quarter Century of UNIX.* Reading, MA: Addison-Wesley, 1994.

Salus, Peter. "The Daemon, the GNU and the Penguin." *Groklaw.* November 1, 2005. Accessed April 26, 2016. http://www.groklaw.net/articlebasic.php?story =20051031235811490.

Stallman, Richard. "The GNU Project." Accessed December 20, 2016. https:// www.gnu.org/gnu/thegnuproject.html.

Stallman, Richard. "Stallman on Qt, the GPL, KDE, and GNOME." *Linux Today.* September 5, 2000. Accessed April 26, 2016. http://www.linuxtoday.com/ developer/2000090500121OPLFKE.

Steed, Judy. "Freedom's Forgotten Prophet." *Toronto Star*, October 9, 2000.

Tanenbaum, Andrew. *Operating Systems: Design and Implementation.* Englewood Cliffs, NJ: Prentice-Hall, 1987.

Tanenbaum, Andrew. "Some Notes on the 'Who Wrote Linux' Kerfuffle, Release 1.5." Accessed April 25, 2016. http://www.cs.vu.nl/~ast/brown.

Tanenbaum, Andrew, Linus Torvalds, et al. "LINUX Is Obsolete." Google Groups. Accessed April 25, 2016. https://groups.google.com/forum/#!topic/comp .os.minix/wlhw16QWltI.

Torvalds, Linus. "Free Minix-like Kernel Sources for 386-AT." Google Groups. October 5, 1991. Accessed April 26, 2016. https://groups.google.com/forum/#!topic/comp.os.minix/4995SivOl9o.

Torvalds, Linus. "LINUX: A Free Unix-386 kernel." October 10, 1991. Accessed April 26, 2016. http://oldlinux.org/Linus/index.html.

Torvalds, Linus. "LINUX's History." July 31, 1992. Accessed April 26, 2016. https://www.cs.cmu.edu/~awb/linux.history.html.

Torvalds, Linus. "Notes for Linux Release 0.01." Accessed May 26, 2016. https://www.kernel.org/pub/linux/kernel/Historic/old-versions/RELNOTES-0.01.

Torvalds, Linus. "Release Notes for Linux v0.12." Accessed May 26, 2016. https://www.kernel.org/pub/linux/kernel/Historic/old-versions/RELNOTES-0.12.

Torvalds, Linus. "What Would You Like to See Most in Minix?" Google Groups. Accessed April 26, 2016. https://groups.google.com/forum/#!topic/comp.os.minix/dlNtH7RRrGA.

Torvalds, Linus, and David Diamond. *Just for Fun: The Story of an Accidental Revolutionary.* New York: HarperBusiness, 2001.

"Validating the Source Code: How One Vendor Does It." *Computer Law and Tax Report*, February 1980.

Van Rossum, Guido. "Origin of BDFL," July 31, 2008. http://www.artima.com/weblogs/viewpost.jsp?thread=235725.

Von Krogh, Georg, and Eric von Hippel. "The Promise of Research on Open Source Software." *Management Science* 52 (2006): 975–983.

Weber, Steven. *The Success of Open Source.* Cambridge, MA: Harvard University Press, 2004.

Weizenbaum, Joseph. *Computer Power and Human Reason: From Judgment to Calculation.* San Francisco: Freeman, 1976.

Williams, Sam. *Free as in Freedom: Richard Stallman's Crusade for Free Software.* Sebastopol, CA: O'Reilly, 2002.

"Wine History." WineHQ. Accessed April 26, 2016. https://wiki.winehq.org/Wine_History.

Wirzenius, Lars. "Linux Anecdotes." April 27, 1998. Accessed April 26, 2016. http://liw.fi/linux-anecdotes.

Wirzenius, Lars. "Linux at 20." Accessed April 26, 2016. http://liw.fi/linux20.

Yood, Charles. "The History of Computing at the Consumption Junction." *IEEE Annals of the History of Computing* 27 (2005): 88.

Young, Robert. "Interview with Linus, the Author of Linux." *Linux Journal*, March 1, 1994.

# Index

GNU Free Documentation License, 268

GNU General Public License. *See* GPL

GNU/Linux distributions, 165–173, 257. *See also* Debian, Gentoo, Red Hat, Slackware, SUSE, Ubuntu, Yggdrasil

  naming of, 153–155, 215–216

"GNU Manifesto," 215

Google, 244

  and Android, 247–251

GPL, 84–85, 88, 89–90, 100, 103–104, 127, 146–151, 154, 158, 174, 179–183, 188, 190, 193, 194, 196, 197, 213, 220, 225–227, 233, 249–251, 262–263, 267

Gsh, 75–76, 82, 106, 185, 202

GTK+, 180

Hacker culture, 6, 37–50, 68, 73, 80, 101, 149, 213–214, 218

  crisis of, 52–57

Hadoop, 196

*Halloween Documents*, 228–232, 233

Harmony (programming library), 180

Hecker, Frank, 245

Henkell-Wallace, David, 80

Hippel, Eric von, 12

"Homesteading the Noosphere," 220

HP-UX, 183

HTML, 189, 225

HTTP, 189–191, 195

Hurd, 92, 96–100, 137, 150, 152, 154, 157, 227

Hybrid code, 236

IBM, 38, 56, 58, 65, 118, 187, 195, 204–206, 222, 234, 238, 244

Icaza, Miguel de, 179

Internet, 10, 22, 40, 42, 57, 59–60, 64, 78, 85, 103, 106, 109, 116–118, 123, 126, 142, 143, 147, 160–161, 165, 168, 174–175, 188–189, 196–197, 199, 200, 203–205, 208, 222, 225, 227, 230, 243, 245, 257, 259–261, 263–267, 270

  open standards on, 188–189

  role in FOSS development, 106, 159–161